Preventing Corporate Fiascos

Thang Nhut Nguyen

Preventing Corporate Fiascos

A Systemic Approach

Thang Nhut Nguyen
California State University,
Long Beach
California, USA

ISBN 978-1-137-48964-7   ISBN 978-1-137-49250-0   (eBook)
DOI 10.1057/978-1-137-49250-0

Library of Congress Control Number: 2016936777

Printed on acid-free paper

This Palgrave Macmillan imprint is published by Springer Nature
The registered company is Nature America Inc. New York

CONTENTS

LIST OF FIGURES

LIST OF TABLES

CHAPTER 1

Prologue

On the What's and Why's

Over the last two decades, corporate fiascos leading to institutional bank-ruptcies have been a great challenge to many: law makers, government watchdog agencies, judges and juries, executives, boards, researchers, pro-fessionals, graduate students, and so on. Numerous investigations have attempted to document what happened and how, and to understand why. In some complex cases the documents and artifacts amount to thousands of pages of: congressional hearings, court records, books, media articles, research papers, and the like—as in the cases of Enron and WorldCom in 2002.

The fiascos were very costly. The impact on their environment was imme-diate. Their ripple effect could last for years or decades. New regulations and domain-specific reforms were proposed and enforced. Subsequent fiascos and bankruptcies kept recurring, however. The collapse of Lehman Brothers in 2008, causing economic turmoil, was an example.

I am driven to address the problem from a systemic perspective. Therefore the book is entitled *Preventing Corporate Fiascos: A Systemic Approach*. This Prologue introduces my view on fiasco prevention and shares the rationale for its development and existence.

Traditional approaches have been on managing or curing fiascos after they have happened (Pfarrer et al. 2008). My systemic approach empha-sizes prevention, rather than manage or cure. I explore four ideas underly-ing my view: (1) *Systemic scope*: fiascos occurring in an institution should

© The Editor(s) (if applicable) and The Author(s) 2016   1
T.N. Nguyen, *Preventing Corporate Fiascos*,
DOI 10.1057/978-1-137-49250-0_1

be seen as being caused by the institution's constituents (e.g., its people), and considered as part of the market and the economy (i.e., a scope much larger than the institution itself); (2) *Signs and symptoms*: should be detected and paid attention to early enough, since they are always there. By the time they surface it is normally too late and bankruptcy will inevitably follow; (3) *Corporate decisions*: should be considered as the key factor leading to a fiasco because decisions drive the institution; and (4) *Control*: who should be involved and how to control both symptoms and decisions.

The first idea of *scope* suggests that the general system theory of von Bertalanffy-Boulding (von Bertalanffy, 1950; Boulding, 1956) could be a suitable *systemic framework*. I argue that the biological spectrum encompassing "protoplasm, cell, organism, community, ecosystem and biosphere" is a good fit in substantiating the institution as community, the market as ecosystem, and the economy as biosphere. The biological spectrum, as a systemic framework, offers an overview of the problem domain within which a solution could exist (i.e., closure property).

The second idea, on the *detection of signs and symptoms*, originates from a rough analogy between *cancer* in humans (and/or any *deadly disease* in general) and *fiascos* in institutions. Cancer is caused by a malignant tumor invading nearby tissue, spreading to other organs or systems. By the time symptoms surface the cancer is already in its later phases: the human with cancer is likely to be facing death (King, 1996).

When an institution is considered analogous to a human, its employees are analogous to the cells. The employees might become "abnormal" and behave as organizational "malignant tumors". They might influence other units in the institution, causing a fiasco. If not prevented, they will potentially lead to bankruptcy. One would want the fiasco symptoms detected early, readily exposed and made transparent to the institution's responsible parties.

We suggest that the cancer analogy can be further investigated so that processes known in one analogue (e.g., cancer in humans) to be applied to another (e.g., fiascos in institutions). The detection of fiasco signs and symptoms will mimic some functionality analogous to the functionality of the human autonomic nervous system on the interstitial fluid and plasma for identifying early invasion. It also mimicks the lymphatic circulatory system in detecting proliferation. The analogues between the components of the biological spectrum offer a rich set of systems thinking available to the pursuit of potential solutions.

The third idea on *decisions* stems from Albert Camus's observation that "Life is a collection of choices" and from Antonio Damasio's suggestion that most decisions are emotion driven (Damasio, 2005). It is this set of decisions by the responsible people in an institution, individually or collectively, which take the institution from point A to point B. Individual or group decisions in an institution should be understood early for remediation.

A few researchers, such as Valerie Stewart (Stewart & Stewart, 1981), have pursued a psychological approach to business management using George Kelly's repertory grid (RG) (Kelly, 1963). They have tried to find the causes of problems from the perspective of the decision makers themselves (i.e., from the model of the world they live in).

I argue that George Kelly's Personal Construct Theory (PCT) and his RG technique in clinical environments can be extended beyond Valerie Stewart's management applications for the understanding of decisions, rational or irrational, with insights from neuroscience (Kavli foundation, 2011) and/or neuroeconomics (Kahneman's Thinking, Fast and Slow) (Kahneman, 2011). This offers an opportunity for finding the root causes of the complex symptoms-decisions problems with assigned or computed criticality values.

With numerical measures of decision criticality, one could then attempt to model both the decisions set and symptoms set as a measurable space in which some basic properties—beyond those in Andrey Kolmogorov's formulation of probability (Kolmogorov, 1956), such as non-commutative property—can be relaxed.

The fourth idea suggests additional *control and governance* with a check and balance capability placed in the capable hands of *an Oversight organization unit*. This organizational unit is parallel to the corporate line of command. This added functionality promotes corporate stability and avoids potential abuse by corporate top executives and management team, as often seen in past fiascos. The four ideas combined give rise to a conceptual model for fiasco prevention.

The above can be further extended beyond fiasco problems in institutions and towards market and economy components. The brain works so well with its networked neurons and supporting glia cells in generating good and bad human thoughts and communicating them via language and speech. We wonder what could tie networked institutions together, and what support analogous to glia cells might help generate and organize creations—such as financial products, markets, business ecosystems, or the economy—to interact with one another for better or for worse.

I postulate that all components must be governed by the laws of nature. The major laws are Newton (force and gravitation), Coulomb (charge), Faraday (induction), Maxwell (electromagnetism), Planck (quantum), and Einstein (relativity). For example, if humans in an economy are considered analogous to particles in a human body, what can possibly be learned from Coulomb laws, Faraday laws and Maxwell laws on electric charges and magnets to address institutional influence, market force, etc. in an economy?

Within institutional, market, and economic environments can decisions be viewed as forces, fields, and energy which could move institutions, markets, and their associated economies from one state to another? Is there anything equivalent to Einstein's theory of specialized or general relativity in the economic space-time environment? Does an economic curvature exist which is similar to the gravitational space-time curvature? These questions make sense since all *cells, humans* (as organism), *institutions* (as community), *markets* (as ecosystem) and *economies* (as biosphere) are all part of the biological spectrum in which the above laws are formulated.

The above summarizes the concepts and processes underlying the current organization of the book its eight chapters, as follows.

Chapter 2, on *corporate fiascos*, reviews and details some selected fiascos of the last two decades. The whats and hows of fiascos are exposed in terms of: (1) *exceptions* (as signs and symptoms of faulty events in terms of what happened where, by whom, how, and why); (2) aberrant *decisions* from the decision maker's perspective; and (3) *control* issues.

Chapter 3 is on the formulation of a *systemic approach* which is based on a biological spectrum. This approach leads to a *systemic framework* in which a *conceptual model for prevention* is sketched. The model for prevention consists of: (1) the *institution* in which fiascos might happen; (2) the domain of *information exceptions* with a focus on early detection; (3) the domain of understanding *emotion-driven decisions* and *decision making*; and (4) the control domain partially and concurrently responsible by an *Oversight organization* unit.

Chapter 4 details corporate *information exceptions management* with a focus on the detection, validation and transparency of *exceptions* in terms of signs and symptoms. The symptoms are always there, even if undetected. They are similar to cancer symptoms, which are hidden below the awareness or consciousness level in the human body, therefore cancer growth and invasion is undetected. The analogy to cancer helps identify features in exceptions management systems analogous to autonomic processes and immunization systems in the human body.

Chapter 5 details the *corporate decisions* in which a modified George Kelly RG is used for understanding, explaining, evaluating and measuring them psychologically. We are particularly interested in irrational decisions, as discussed in Daniel Kahneman's Thinking: Fast and Slow, and in Descartes' Error from Antonio Damasio. This is an addition to current well-established approaches to probability-based decision analysis and/or quantitative methods. This takes into account the neuroscience or neuroeconomics decision approach.

Chapter 6 details an *organization* unit for *enhanced management control* and *corporate governance*, looking at the intertwined *exception-decision complex*. The unit is charged by the board with responsibility, authority and accountability for finding exceptions and questioning corporate decisions made by the line of command. The unit operates in parallel with the institution's line of command.

Finally, *Chapter* 7 exploits the general *extension* of a concept of homeostasis in humans to stability in institutions, equilibrium in markets and balance in economies. We argue that the model is suitable for addressing chaos in the market and turmoil in the economy.

ON THE HOW'S

How did I get here? Initially, I gave a conference paper on the understanding of corporate decisions, co-authored with Khanh P. Tran my research assistant, published in the 2014 Proceedings of Western DSI. Somehow, it caught the eye of Casie Vogel, an associate editor of Palgrave Macmillan. I have Casie to thank for injecting the idea of writing a book. That was how the conference paper and the publishing idea started to fuse as a zygote.

Over a few weeks following April 2014 the book's *placenta formation,* analogous to the embryonic development in a pregnant human, took place in the communication between Palgrave Macmillan (including anonymous reviewers) and me on the development of a prospectus. It became a contract in June. The manuscript took the form of an *embryo.*

Next, it entered the *differentiation* phase. Just like any embryo which develops from three layers: *ectoderm, endoderm,* and *mesoderm,* the manuscript split over the months from June 2014 into three major parts. The ectoderm (forming the skin, the nerve tissues and spinal cord) became the *framework*. The endoderm (forming all linings of the organs) was the *model*. The manuscript's mesoderm (forming the backbone) became the detailed *table of contents*. The mesoderm, with all of its *somites* developing into organs and organ systems, led to detailed chapters. I have been engaged in developing a system theory on institution, market and economy for a long time. I restricted it, however, to one area of application: prevention of corporate fiascos. The manuscript embryo has developed into a newborn, empirical theory, with its first breath of life being the production of this book.

To build this model for prevention, at least empirically, I relied on past fiasco cases for information. I use analogies and analogical arguments as started by other theorists. Two important writings, among others, have influenced my theoretical development research method. First was the *analogy and analogy reasoning* article by Paul Bartha, and second, posts on the Academy of Management Review (AMR) website on theory development. The latter were suggested by Professor Mike Pfarrer, an associate editor of AMR. I have exchanged emails and received valuable comments on an article submitted to AMR from Mike Pfarrer and his review team. The submission to AMR was initially suggested by Professor Tom Stafford, the editor-in-chief of *Decision Science Journal* after his initial review of my original article. He offered excellent comments. I have to thank these editors.

My academic background had something to do with the topic addressed in the manuscript, although it appears irrelevant at first. I graduated with a BSc in Electrical Engineering from Université Laval, Quebec, Canada. It was followed by an MSc in Information and Computer Science from the Georgia Institute of Technology, with some MS/OR knowledge, and a Ph.D. in Information Technology and Engineering from George Mason University, VA. All the above credentials followed the Certificat d'Études Supérieures en Mathématiques Generales from University of Hue in (former) South Vietnam. I have the Colombo Plan, USAID and the former government of South Vietnam to thank, since they provided the scholarships for me to attend those universities.

During all these years of schooling, my family made a lot of sacrifices for my academic advancement. I have my parents, Nguyen Thuc Tam (d) and Truong Thi Hong Quang (d), my parents-in-law, Chau van Thanh (d) and Tang Ton Nu Cam Van (d), and all members of my own family (my wife, Chau Thi Bich and our children, Nguyen Nhut Vu Anh (d), Nguyen Nhut Van Uyen and her husband John-Paul Napoles, and Nguyen Nhut

Quoc Anh) to thank and apologize to, since I have neglected them at times as a son, a husband, and a father. I would also like to thank Professor Harry Stephanou (d), my former thesis advisor, who introduced me to the world of robotics and automation, and to academic research.

My business experience has been with IBM and the former Candle Corporation now IBM Tivoli (1980s and 1990s), SAIC (2000s), and other organizations prior to the 1980s, such as the American Bankers Association (ABA), Value Systems Engineering, US Chamber of Commerce, and Litton Computer Services of the former Litton Industries. Each has been a wonderful working experience to sharpen my skills, mostly in the areas of computer systems, systems engineering, and applications. These included systems requirement development, design, implementation, production, and maintenance of software, and business insights. I began studying decision modeling after experiencing some major failures of software development (IBM product failure, US Army Future Combat System development, and others) at the places where I was employed. I have my former managers and co-workers in these institutions, and their clients, to thank for these opportunities.

The disciplines and prior fragments of experience I have been involved in did not indicate that someday I would be interested in systems theory. It was during my tenure at CSULB that systems thinking started to seed in my mind around the mid-2000s, after I read the general systems theories (GST) of Ludwig von Bertalanffy and Kennett Boulding, but it was all forgotten. I had no need to use it then.

Then my researches led to human intelligence and decision making, and decision making for modeling. So I started to read neuroscience. I began with Eric Kandel, Steven Pinker, and others. I tried "true" software intelligence in a different direction to the traditional AI. I looked into how to build a software neocortex which can learn like a human newborn. I first attempted to explore software learning via vision, since that is the best known of the five senses. The basic question was "*what does a visual memory look like in human long-term memory?*" I explored mimicking the human vision process from retina to short-term memory (no one knows what a seen object looks like in long-term memory after it is translated or transformed from short-term or working memory by the hippocampus).

With my only collaborator and contributor, Tony Phan, a molecular biologist and assistant professor at the University of Western Australia, I hypothesized a visual memory for storage in and recall from long-term memory as multi-layered unions of lines and filled-in surfaces. The project started well with a couple of conference papers and tutorials on biologically inspired systems at the IEEE international conference in 2008 with Tony Phan. Tony however died of motor neurone disease (MND) at the age of 32. I had to put this line of research on the backburner. My thanks go to Tony Phan who introduced me to the world of biology.

Phan's departure coincided with some of the largest bankruptcies reported in the news and media during the late 2000s. Abundant literature on different fiascos and bankruptcies was readily available in extensive congressional hearings, academic research and professional investigations (legal, financial, accounting, organizational, managerial, ethics, etc.). It was determined that most were caused, arguably, by humans exercising fraud and making aberrant decisions. I began to explore diseases in humans due to gene mutation and/or viruses in host cells. I wanted to use insights from diseases in humans in the, so-called, diseases in institutions, such as fiascos leading to bankruptcy.

I realized that human decision makers are part of the biological spectrum as much as the *protoplasm*, the *cells* making up the *human*, the *institution*, the *market* and the *economy*. I was led to look into George Kelly's PCT and RG techniques to understand human decisions and decision making in an institution. This includes considering Paul MaClean's triune model of the brain and Kahneman's Thinking, Fast and Slow, and others. I began to investigate corporate decisions with a focus on irrational decisions within the context of fiasco prevention from the biological spectrum's view.

My intention is to eventually look at the set of corporate decisions D as part of a measurable space $\{D, \mathbf{D} \text{ and } \mu_D\}$ where $\mathbf{D}$ is the σ-algebra on D, and μ_D is a subjective measure of criticality of decisions. The criticality measure does not have to observe probability properties, such as commutative property, since real-life emotion-driven decisions are not commutative. Using μ_D assigned RG technique, my objective is to look at the generated $\sigma(\Delta)$ where Δ is a subset of D, for the identification and selection of an optimum decision to remediate exceptions prior to the potential fiasco.

At first I saw no systemic implication of Kelly's PCT to business decision problem solving until I came across Antonio Damasio's article on Descartes' Error, which states that most decisions are emotionally driven. The biological spectrum and psychological PCT became the *systemic foundation* of my proposed framework. As such, I thought it would be extendable to markets and the economy since both are human-driven and are also part of the biological spectrum.

Finally, I have many friends and colleagues to thank for making this book a reality, in alphabetical order, Professor Lori Brown, Dr. Khiem Cai, Michelle Nguyen, Richie Nguyen, Sommer Nguyen, Dr. Patrick O'Rourke, and Professor Mike Walter. They have helped review the draft version and the blurbs on the jacket. Last but not least, my thanks go to Khanh Tran, my longtime friend and research assistant, and to Stacy Noto, Palgrave Macmillan editor, Marcus Ballenger, editorial assistant, and other staff members of Palgrave Macmillan and later of Springer Nature who have been involved in this project, for their incredible patience and publishing assistance, and valuable time.

Corporate Fiascos

Dozens of corporate fiascos leading to bankruptcies over the previous two decades have been subject to numerous investigations and research. Each case was complex. The cost of each fiasco or bankruptcy was huge. Their impact immediate. The ripple effects long-term. A few selected fiascos are used in this chapter for illustration and discussion based on information from the literature.

We witnessed the collapse of Baring Bank in February, 1995; the bankruptcy of Lehman Brothers in December, 2008; and others in between, such as WorldCom, Adelphia Communications, Tyco International, Parmelat, and so on (Brickley, 2003; Gale, 2012; Heller, 2003). Many were the result of sophisticated frauds but their root causes were found to be different. Table 2.1 summarizes the whys and hows of four typical bankruptcies caused by fraud.

In general terms, Barings Bank collapsed due to the work of one manager, Nicholas Leeson. The main root cause was, allegedly, his uncontrollable self-interest. The conflicting dual role of Leeson in both front and back office operations was a critical issue that was ignored by Barings executives. The collapse was also caused by the absence and/or negligence of management control over the bank's operations (Rawnsley, 1995).

In the other three cases shown in Table 2.1, the bankruptcy was essentially due to a group of executives. For example, the greed exercised by two main Enron executives, CEO Jeff Skilling and CFO Andrew Fastow, was particularly strong. The dual role of Arthur Andersen Accounting as both accounting auditor and consultant at Enron should not have been

© The Editor(s) (if applicable) and The Author(s) 2016
T.N. Nguyen, *Preventing Corporate Fiascos*,
DOI 10.1057/978-1-137-49250-0_2

Table 2.1   Typical collapses caused by fraud

When	Institution: who and what	Assets	Why	How
Feb 1995	**Barings Bank** **Who:** Nicholas Leeson **What happened and when** Bankrupted when liability exceeded £830M	£630M	Manager's self-interest Management control negligence Internal audit Dual role of Leeson: front and back offices Fraud	Use of error account 88888 Hid losses, borrowed from client floating funds, own commission, cut positions Doubling scheme
(Leeson, 1996; Rawnsley, 1995)				
Dec 2001	**Enron Corporation** **Who:** Jeff Skilling, Andrew Fastow **What happened and when** Bankrupted after first reconsolidation of 2001 Q3 followed by reconsolidation of 1999–2001 financial statements	$65.5B	Executive greed Corporate governance Enron personnel review system Faulty organization design Arthur Andersen both auditing and consulting Accounting fraud	Use of special-purpose entities (SPEs) Transformed cash flow from financing into cash flow operations Nonconsolidated losses
(Rapoport, Van Niel and Dharan, 2009; Sridharan et al. 2002; Thomas, 2002)				
Jul 2002	**WorldCom** **Who:** Bernard Ebbers, Scott Sullivan **What happened and when** Bankrupted when accounting fraud discovered Improperly accounted for more than $3.8B in expenses	$107B	Executive greed Executive self-interest Corporate governance Corporate culture Internal control Auditing firm failed to act Accounting fraud	Merger and acquisition without consolidation Expenses as capital expenditures $400M loan to Ebbers
(Belson, 2005; Lyke & Jicking, 2002; Thornburgh, 2004)				

(*continued*)

Table 2.1 (continued)

When	Institution: who and what	Assets	Why	How
Sep 2008	**Lehman Brothers** **Who:** Richard Fuld Jr., Erin Callan, other executives **What happened and when** Bankrupted when liability was in the order of assets, with only $25B in capital	$691B	Executive greed Corporate governance Lack of risk management Downturn of subprime market: collateralized debt obligations (CDO) and credit default swaps (CDS) Leverage, cash flow Executive incentives Accounting fraud	Borrowed cash from money market fund High-risk, high leverage strategy Use of Repo 105 to reduce debts Non-disclosed losses Negotiations with Barclays Bank and Bank of America failed FED refused to bail out

(D'Arcy, 2009; Estrada, 2011; Le Maux & Morin, 2011; Michel 2013)

allowed. One learned later that the fraud was masterminded by top executives via hidden losses and faked gains in the form of complex special-purpose entities (SPEs).

Following the fall of Enron and WorldCom in 2002, the Sarbanes-Oxley Act (Sox, 2002) was enacted in the same year to tighten up the responsibility and accountability of CEOs and CFOs. There were also reforms in accounting practice and other domains. One would think that executives could not or would not repeat the same fraud. However, Lehman Brothers filed for bankruptcy 6 years later, in 2008 (Azadinamin, 2012). Lehman Brothers executives used an accounting vehicle similar to those used by Andrew Fastow of Enron (i.e., SPEs) to exercise fraud. Known as Repo 105 they hedged funds and moved expenses off the books. Lehman Brothers also lied about the funding of $500M from Bank of America.

In the above cases (involving either a single manager or a group of executives) there were also a lot of disconnects between management, top executives, and their boards. In Barings, the top executive was unaware of what Nicholas Leeson was truly doing. In Enron, the Board of Directors was unaware of what the Enron executives were up to. In the case of

Lehman Brothers, CEO Richard Fuld Jr. ignored all recommendations from their top-notch managing directors and professionals.

We observe that a fiasco leading to bankruptcy is the result of a series of decisions made by the responsible parties. In the Barings case, an initial decision was made by Leeson to use the error account named 88888 for recording daily discrepancies in transactions. Use of an error account was a common practice in banking, however Leeson used it to conceal losses. The fraud lasted for some time. In July 1993 Leeson was able to clean up the fraud and balance out the losses. He could have stopped then. But he had to continue when newer losses occurred.

In an attempt to recover Leeson made the decision to trade derivatives using a high-risk scheme called doubling strategy, commonly known in gambling circles. Trading included the Nikkei 225 index, Japanese Development Bonds (JDB), and Euroyen. Leeson explained how he did it in his Rogue Trader book published in 1996. The final loss amounted to a liability over and above Barings' assets (Leeson, 1996, 2012).

The Enron case involved Jeff Skilling's decision to extend a mark-to-market strategy to other businesses, and Andrew Fastow's decision to hedge via special purpose entities (SPEs). Both initially worked successfully in the effort to make Enron a gas bank. Its first project called JEDI used hedge funds from CALPERS (CA employees' pension system) (Jickling, 2003).

Enron then expanded to other industries using an asset-light strategy rather than a heavy-asset strategy (Chatterjee et al., 2002). Enron needed huge hedge funds, so the malicious use of SPEs was in the thousands. They were used with the intention of keeping expenses off the books and to cook up financial statements. It was found that a reconsolidation was required after the 3Q 2001 statement. This led to financial re-statements for the three prior years (1997–2000). Enron stock prices plunged from around $90 to around 20 cents.

In WorldCom's case, executives decided to expand growth by mergers and acquisitions. Facing huge costs, WorldCom exercised accounting fraud in reporting expenses as capital expenditures (Lyke et al., 2002). With Lehman Brothers, the executive decision was to enter the subprime lending market with a high-risk strategy, which eventually incurred huge debts. Consequently, CEO Richard Fuld had to use Repo 105 to conceal losses (Michel, 2013).

Lacking cash, Fuld thought there would be buyers or rescuers. Negotiations failed with potential buyers, Korean Development Bank,

Barclays, and Bank of America. The Federal Reserve Bank of New York decided not to bail them out.

In terms of impacts and the ripple effect: the Enron fiasco caused around 5000 job losses on the day of its bankruptcy filing; some 20,000 former employees and retirees suffered $1B loss to their pensions and 401(k) plans; investors underwent $50 billion in total losses; court costs, official and undocumented, were estimated to amount to $1B; Enron's accounting audit and consulting firm, Arthur Andersen, subsequently collapsed; and the impact was felt in many other areas of business (Paulsen, 2002).

In 2008, Lehman Brothers became the biggest fiasco of the decade. Liabilities were around $619 billion and assets $629 billion at the time of filing. Some 15,000 employees lost their jobs and some 25,000 lost most of their stock. The downturn in the subprime market was one of the main causes that led Lehman Brothers to make inappropriate decisions on Repo 105 and commit other frauds.

The immediate impact of Lehman Brothers' collapse was considered by some authors as analogous to an earthquake in the investment market. The domino effect began on the filing day, September 15, 2008. It led to the collapse of General Motors in June 2009 and the fiasco of Ernst & Young in 2010. The effect on secured credit market and the economic implications have been well documented (IESE Professors, 2013).

There were also fiascos and bankruptcies caused not by fraud but by: external failures; or bad internal strategies, risky mergers and acquisitions; aberrant decisions resulting in losses and a lack of liquidity. These included Pacific Gas and Electric Co., Global Crossing, Conseco, Delta Airlines, Washington Mutual, Thornburg Mortgage, and General Motors (see Table 2.2).

Also worth mentioning are governmental fiascos, particularly those in the US Armed Forces: the US Army Future Combat System (FCS) cancellation in 2009; followed by the US Air Force Expeditionary Combat Support System (ECSS); and US Marine Corps Global Combat Support System (GCSS). These fiascos caused hundreds of billions of dollars of wasted funding. Recently the turmoil caused by Healthcare.gov also made the news (GAO, 2009; GAO, 2012; Schadler, 2013).

The Pacific Gas and Electric Company (PG&E) did not commit any fraud; its collapse was as a victim of a scam by Enron. The California electricity crisis began when Enron manipulated the energy market. Enron took advantage of a deregulation bill in 1996 and created a spot market in which suppliers such as PG&E had to buy electricity at a much higher price to satisfy client demand, which had been set at much lower fixed prices. Exercising the gas bank strategy, Enron created a shortage in energy in California by shutting down the pipeline and causing blackouts multiple times to raise prices. The congressional hearings, court records, and other research contain detailed accounts of this scam (Congressional Testimony-1, 2002; Congressional Testimony-2, 2002).

See other cases cited in Table 2.2 for a summary of why those institutions collapsed.

BARINGS BANKRUPTCY CASE STUDY

This case is compiled from numerous sources: newspapers, reports, books, journal articles, and so on. We take the view that corporate fiascos result from a series of faulty events, intertwined with aberrant decisions that have either caused or followed those events. The faulty events are called exceptions, which have preceded and/or followed some managerial, or operational decision that turned aberrant. In the following, we organize the happenings relevant to this case from three perspectives: (1) exceptions, (2) decisions, and (3) control.

What Happened and When?

Barings Bank collapsed on February 26, 1995 (Leeson, 1996, 2012). The key employee responsible for the collapse was Nicholas Leeson. He single-handedly managed to incur debts of £827M while bank assets were £630M. Given the facts that have since been uncovered, it was a truly unbelievable fiasco.

Leeson was employed by Barings Settlement Division in July 10, 1989 to work on futures and options. He was assigned to the Jakarta office to sort out share certificates worth around £100M. These were considered as Barings liabilities. After 10 months, he was able to reduce Barings exposure to £10M. He returned to London in March 1991 a hero.

Leeson then traveled extensively with Tony Dickel of Barings to explore new opportunities around the world. One of their recommendations was to boost Barings' Singapore office. In October, 1991 Leeson was offered the job of running Barings Singapore, as an obvious choice to Barings executives.

Leeson arrived in spring 1992 as General Manager of Barings Singapore. He hired traders for futures and options on three equity markets: (1)

Table 2.2 Other collapses without frauds

When	Who and what	Assets	Why	How
Apr 2001	**Pacific Gas and Electric Co.** **What happened:** Bought power at fluctuating prices to sell at fixed prices (Holson, 2001)	$36.1B	Deregulation State help too slow	Sold natural gas power plants, under capacity
Jan 2002	**Global Crossing** **What happened:** High debts (The Economist, 2002)	$30.1B	Too slow to meet market demand	Failed fiber-optic strategy
Dec 2002	**Conseco** **What happened:** Huge debts (Norris & Berenson, 2002)	$61B	Massive debts and loans to officers	Merger and acquisition
Sep 2005	**DELTA** **What happened:** Debts (Isidore, 2005)	$21.8B	9/11, fuel prices, Katrina	Too slow to trim costs
Dec 2005	**CALPINE Corporation** **What happened:** Debts (Becker & Polson, 2005)	$27.2B	Debts due to loans and bonds	Lost fight with bond holders
Apr 2007	**New Century** **What happened:** subprime crisis (Stempel, 2007)	$26.1B	Improper and imprudent practice	KPMG audit negligence
July 2008	**IndyMac Bank** **What happened:** Subprime mortgage crisis (Reuters, 2008)	$32.7B	Large scale NINJA loans	NINJA loans defaulted
Sep 2008	**Washington Mutual** **What happened:** CDOs (Grind, 2012, AP 2008)	$328B	Massive withdrawals by customers	FDIC control
June 2009	**General Motors** **What happened:** Huge costs (Hill, 2010)	$91B	Terms of government loans	Government stepped in too late

Nikkei 225 stock index (similar to the US Dow Jones); (2) 10-year Japanese Development Bonds (JGB); and (3) Euroyen between OSAKA Exchange or OSE of Japan and Singapore International Money Exchange or SIMEX of Singapore. He activated Barings' seat on SIMEX and went on the SIMEX floor to assist traders. After he passed the exam to become a trader, he started to perform arbitrage activities himself on futures and options contracts.

Futures contracts, or *futures*, are derivatives contracted between a buyer and a seller. A contract allows the seller to sell an underlying asset at a future date and the buyer to purchase the asset on that date or before. Another form of contract is called *options*, which is different from futures in that the buyer has no obligation to buy the asset.

These contracts required buyers to post an initial margin as a fixed and small percentage of the overall value of the contract, about 5 % or less in cash or securities. SIMEX held the margins in a separate account and the contracts were priced daily. The margin account had to be maintained to the level set by SIMEX at the end of day. Winners would be paid from the losers' margins. Funding for a margin account came from clients or from three Barings companies involved in this trading: Barings Securities, Barings Securities London, and Barings Securities Japan.

In an option the spike price of the asset was the price specified in the contract. The buyer could purchase the asset before the expiration date, sell it, or wait until the expiration date. The gain or loss was the difference between the spike price and the asset price.

Leeson and his traders exercised futures and options as follows. In Tokyo, a Barings trader would watch the Nikkei 225 index and pass it along to a Barings trader in Singapore. Barings Futures Singapore would buy at a lower price on SIMEX and immediately sell them through a Barings trader in OSE for a slightly higher price. Depending upon the volume, a difference of 10 points in the index could earn tens of thousands of pounds. This had to happen within seconds through a coordination on the floor and via telephone communications.

Three major events helped Leeson in Barings' futures trading business. First, there was a change of regulations in OSE in Japan which drove investors to SIMEX in Singapore for lower premiums and interest. As a result, trade volume went from 4000 to 20,000 a day.

Second, Barings Settlements London advised Leeson to keep detailed errors in the Singapore office for end of day consolidation rather than report them to the UK headquarters. Leeson was asked only to report margins. Leeson created an error account called 88888 (or five 8s) for end of day consolidation, which gave him the opportunity to hide the details.

The third was that, via Ang Swee Tian, President of SIMEX, Leeson was introduced to a new client, Phillip Bonnefoy who traded in large volume, roughly 4000–5000 a day. Bonnefoy became the largest trader on SIMEX.

In July 1992, one of Leeson's new employees, Kim Wong, made a mistake on the SIMEX floor. She sold rather than bought 20 options contracts worth around £20K. Leeson reported this to his Singapore manager, Simon Jones, who suggested forwarding the problem to Barings Futures Tokyo for help. Leeson decided not to. Instead, he decided to blank out the loss by creating a fictitious client and concealing it in error account 88888. Leeson was able to do this because he had two hats: as general manager for settlements in the back office; and as a trader on the trading floor.

Three days later, the market rose 200 points. The loss now went up to £60K. It was too big and too late to report to Simon Jones or the Tokyo office, especially since Leeson had not done what Jones had asked earlier. Leeson continued to hide the loss in that account.

A further incident was due to George Seow, who had carried out a lot of trading and made many errors. On one occasion he had bought rather than sold 100 contracts worth around £8M with a margin of around £150K. To this date, Leeson was left with 420 contracts he could not sell, although he told his clients they were sold. Again, he concealed the loss in the error account.

Leeson's core trading from the end of 1992 was selling put and call options on the Nikkei 225 index in straddles. This provided Leeson with cash for premium payments to SIMEX. No payment needed to be made to the buyers of the options until the end of the contracts. Thus, as long as the Nikkei index remained within a reasonable range, Leeson's scheme was in no danger of being discovered since he reported the sales as profits.

Leeson devised a scheme to sell options for the right amount in Japanese yen to pay SIMEX and to ask for cash in dollars from Barings Settlements London. He was able to show, despite all the losses, zero on the balance sheet and on the profit and loss statement.

Unfortunately, in addition to the losses made, the market went against Leeson. By the end of 1992 the loss recorded in error account 88888 was £2M. Luckily, in July 1993 the account balanced out. However, by the end of 1993 the loss was accumulated to £23M. Six months later, in June

1994, it was £116M. By December of 1994, it amounted to £208M. The loss exceeded the profit reported at £205M before tax and before £102M bonus. On February 27, 1995, the loss was £827M.

How the Collapse Finally Happened

The Nikkei index between March and December 1994 was close to an 8-year low and fluctuated between 19,000 and 21,000. During 1993 Leeson managed to sell options for cash to pay the premium to return the balance to zero. At the end of 1994, however, it was different. Error account 88888 contained 1000 March 1995 futures contracts. Leeson was 7.78B yen short (or £50M). He had made an entry in the ledger as if Barings was owed it by SIMEX from a third party. He identified the third party as a fictitious Spear Leeds & Kellogg (SLK), Leeson knew he was in trouble.

Leeson planned to flee during the Christmas holidays but went back to Singapore on January 6, 1995. In his book *Rogue Trader*, Nicholas Leeson described his activities and the reasons why he came back from the beginning of January 1995 up to the day he fled to Indonesia. He divided this into three periods: the first covered January to February 6; the second from February 6 to 17; and the third covered the week of February 17 to 23 (Leeson, 1996, 2012).

From January to February 6 From the first week of January, Leeson's scheme was to buy many more futures to swing the market. He needed some $10M a day from Baring Settlements London for the margins. At times, he asked for $30M or $40M a day.

The KOBE earthquake struck Japan on January 17, 1995. As a result, the Nikkei index fell from 19,350 to 18,950 on January 20, 1995. It fell again to 17,785 from 19,241. Leeson had a big hole of 7.78B yen (equivalent to £50M) in his error account. This amount was fraudulently reported as income. Two additional incidents occurred during this period (Brown et al. 2008).

First, SIMEX discovered that Barings had financed the trading margins rather than clients. It was a violation of SIMEX rules. This was stated in a letter from SIMEX to Simon Jones on January 11, 1995.

Second, Barings Internal Audit, performed by Coopers & Lybrand, asked about the 7.78B yen reported as income, which they were unable to trace. Barings executives asked Leeson to account for the untraceable amount of 7.78B yen. These included: Simon Jones (Leeson's boss); Brenda Granger of the Settlements Department, London; Mary Walz of the Financial Product Group; Ron Baker (Mary Walz's boss); Tom Hawes (the Group Treasurer); and Rachel Yong (the Financial Controller).

Leeson started to lie to buy time and then decided to forge a note from Richard Hogen, Managing Director of SLK, to confirm that the balance of 7.78B Japanese yen would be paid on February 2. To complete the lie, he forged a note from Ron Baker of Barings to show that Barings had received the payment. On February 3 Coopers & Lybrand, the internal auditors, cleared the audit report. However, no one saw that money was paid to Barings. Leeson explained that the reason for not receiving the 7.78B yen was an accounting screw-up—a computer glitch.

From February 6 to 17 February 6 was a good day since the market was in Leeson's favor. He closed out 1100 contracts and got £15M. But the loss was still in the order of £200M. He needed more money so he worked on JDB futures and doubled up both JDB and the Nikkei 225. Unfortunately, this amounted to a total loss of £300M or about two-thirds of Barings entire share capital base. During the following weeks, he bought 15,000 Nikkei futures contracts, 55,399 March contracts, and 5640 June contracts. The loss from these contracts amounted to £384M, or 59B yen, as of February 23, 1995. This yielded a liability much higher than the bank's assets.

The Last Three Days: From February 20 to 23 Leeson kept buying in the hope that the price would go up. Somehow, Leeson was still able to convince Tony Railton of Barings to authorize another $30M on February 22 and $40M on February 23 for margin payment. A few minutes before Leeson left his office for good on February 23. The error account 88888 recorded at Nikkei's close at 17,885 Leeson was "*long of 61,039 future contracts, short of 26,000 JDB contracts, and a mixture of Euroyen and Nikkei options*". The situation was devastating (Rawnsley, 1995).

Why It Happened?

A lot has been written on the reasons why Barings ended up bankrupt, both speculatively and factually. Leeson admitted in his book that he took advantage of Barings Bank's new strategy in the securities market, and the negligence in management control. Others said it was: the lack of risk management; Leeson's dual role with front and back office responsibilities; the KOBE earthquake: and other things. Collectively, they all led up to the fiasco.

Leeson had panicked in dealing with the fall of the Nikkei index after the KOBE earthquake. He had no other way out but to use a *double or nothing* scheme. Leeson appeared to have completely lost his mind by the end. He believed he could balance out to zero the error account, as he had done in July 1993. To clear the 7.78B yen for the end-of-year audit and statement by Coopers & Lybrand, he had forged two notes, one from SLK and the other from Ron Baker of Baring Settlements London.

In summary, we might say that Leeson was a capable man and lucky at the start. He was hired to do settlements of futures and options. Within a year, he had reduced liabilities of £100M to £10M in Indonesia. He had a chance to travel the world with Tony Dickel for Barings opportunities. To Barings' management, he was the only and obvious choice to head Barings Futures Singapore.

Circumstances surrounding Leeson in Singapore allowed him to both behave and misbehave. On the good side there was: (1) the creation of the error account for good banking practice (consolidation of errors daily); (2) the Osaka revised regulation boosted the SIMEX business; (3) the arrival of a new client, Phillip Bonnefoy, with a large volume of trading so Leeson could make daily requests for funding from $10K to $40K a day. These requests were accommodated with few doubts or questions from Barings.

On the bad side: the error account gave Leeson a way of hiding loss; his dual role in both settlements and trading offered Leeson the opportunity to exercise fraud; in addition to the management control negligence, the line of reporting was unclear so his trading activities were unsupervised.

It was recognized that Leeson was not in it for the money. He could have walked away with millions of dollars when he discovered the unclaimed shares during his time making settlements in the Jakarta office, but he did not. His dream was to play the derivatives game on the trading floor. He was a hard worker. He took care of his employees' mistakes besides his own. Kim Wong's mistake of £20K through inexperience and George Seow's £50K mistake in trading were both covered in the error account. Was Leeson a caring man or a sick man? Was he selfish or generous? Was he intelligent or just a risk seeker?

There have been a couple of interesting studies on the bank collapse from a perspective other than business. One analysis was done by M. Stein in 2000 (Stein, 2000). Stein looked at the special situation Leeson was in and thought the risk-takers were actually Barings executives and management acting as a *shadow* of Leeson. Therefore, Barings top management tolerated Leeson's behavior. The whole bank considered him a savior who brought wealth in a fashion similar to Christopher Heath, CEO of Barings Bank in the 1980s, who was successful in trading on the Nikkei index.

Another interesting analysis was done by Ian Greener published in Organization in 2006. Greener used Mike Callon's *sociology of translation* scheme, published in 1986, to find answers to Leeson's behavior. Greener built Leeson's *actor-network* by applying the four phases of Callon's scheme.

In its first *problematization* phase, Leeson was identified with his actors in the actor network: his employees, his traders, his superiors, his clients, and SIMEX. In the second phase, called *interessement*, Leeson built a unique and powerful position with support from his loyal staff. He had direct access to Barings management. He was close to and well respected by the SIMEX authority. In the third phase, *enrollment*, he got them enrolled, which made things easier for him. In addition, Gordon Bowser, Head of Futures Settlements of Barings Securities, asked him not to send trading details to London, just margin payment requirements. This allowed Leeson to hide details in error account 88888. In the fourth phase of *mobilization*, he became the authority on futures in Singapore on behalf of Barings, not just on settlements in the back office.

Another analysis was that of Deborah W. Gregory, in *Unmasking Financial Psychopaths: Inside the Minds of Investors*. She attempted to explain Leeson's behavior using Hare's Psychopathic checklist. It appears that Leeson fit the deceitful and grandiose qualities, and some degree of impulsivity, but not as a psychopath. According to Gregory, Leeson was a narcissist who happened to be the right person (young, willing, risk seeking) at the right time and in the right place (Singapore and Barings).

The Bank had been warned several times, as we showed earlier. The first was about the dual role of Leeson raised by James Bax, which was ignored by executives. The second was the report on risk by James Baker and Ian Manson, the internal auditors. The third was from George Maclean's memo to many Barings heads in September 1994 after the margin exceeded the 25 % imposed by the Bank of England. Yet, Leesons' demands for premiums were never truly questioned, beyond some requests from headquarters for verbal explanations. The 1994 year-end audit was easily passed based on the two forged notes.

Despite his charm and good luck Leeson experienced a period of unlucky business towards the end. He seldom won any trading activities based on his knowledge and experience in futures and options. In his first period, to end of 1992, the loss was already in the order of £2M, due to his lack of knowledge, as demonstrated in his handling of trade on behalf of Philipp Bonnefoy.

In the Barings case, we can identify three issues:

1. *Exception issue*: No one except Leeson and his female staff in the back office knew about the existence of the error account until it was revealed by SIMEX in its letter to Simon Jones on June 21, 1993 concerning the 7.78B yen. No prior red flags had been reported, only warnings in passing. No Barings executive paid attention to the details. Even the President of the bank, Peter Norris, casted no doubt on the huge gains from arbitrage activities. James Bax, Regional Manager of Barings South Asia, was the first one who warned about Leeson's role. His warning was ignored. Other warnings came from the audit conducted by James Baker and Ian Manson. One of them was the need for a full-time Risk and Compliance Officer, a warning not seen as justifiable according to top management. The recommendation to place a cap on daily positions (i.e., 200 Nikkei futures, 100 JDBs, and 500 Euroyen futures), has never been observed. As of September 1993, Leeson already held a total of 20,000 futures and options, including 5000 Nikkei futures, 2000 JDB futures, 1000 Euroyen futures, and the rest were options. The exceeded numbers were unbelievable, but mo one in the line of command above Leeson knew this. There were information disconnects, Leeson's abuse and manipulation, and negligence by Barings top management.

2. *Decision issue*: The only decision maker was Leeson, apparently. He coordinated through an actor-network, as described by Callon's *sociology of translation* scheme. He came up with all sorts of maneuvers to request tens of millions each day. Surprisingly most requests were accommodated by Barings London and Tokyo until the last day before his escape.

3. *Control issue*: Leeson reported to many supervisors: Simon Jones in Singapore; Mary Walz and Ron Baker in London; Mike Killian in Tokyo. Leeson actually managed his superiors. No one in the Barings chain of command seriously questioned his activities. He had *carte*

blanche to do anything he wanted. The question was, despite all warnings and decisions regarding Leeson, what was Barings top management thinking?

LEHMAN BROTHERS BANKRUPTCY CASE STUDY

This is one of the most complicated and absurd cases. Again, we organize the relevant facts and thoughts from three perspectives: (1) exceptions detailed in *"What happened and when"* and in *"How it finally happened"*, (2) decisions in *"Why it happened and who did it"*, and (3) control. The Lehman Brothers story was different in many ways from the Barings Bank story.

What Happened and When?

Lehman Brothers filed for bankruptcy on September 15, 2008 primarily due to a lack of cash (Azadinamin, 2012). Also, at the time of collapse the liabilities caused by a high leverage index were in the order of assets around $691B, with only $25B in capital. The decision to file for bankruptcy was made after the Federal Reserve Bank of New York (Fed) refused to bail out the bank. Strangely enough, the Fed had decided to bail out Bear Stearns 6 months before and AIG only 1 day after. The Fed said it had no authority to bail out Lehman Brothers.

The problem started years earlier, around the mid-1900s, when Lehman Brothers decided to enter the subprime mortgage market, as reported by H. Michael in 2007. Lehman Brothers' decision was initiated by a major change to the mortgage market as a result of the *Gramm-Leach-Bliley* Act. This Act allowed Wall Street investment banks to get involved in the mortgage market, which had previously been handled only by commercial banks.

Lehman Brothers considered funding the First Alliance Mortgage Co. by sending out one of its VPs, Erick Hibbert, to investigate the possibility of entering the lucrative subprime market after the *Gramm-Leach-Bliley* Act. Hibbert reported that First Alliance operated as a financial "sweat shop" and that "ethics" was not observed. However, the top executives ignored that fact and concluded that First Alliance had not yet broken any laws. Therefore, the bank funded $500M to First Alliance Mortgage Co. It also helped sell $700M in bonds backed by customer loans. The bonds were called mortgage backed securities, or MBS (Michel, 2013).

The way MBS works is as follows. Mortgage loans are handled by mortgage brokers as the middlemen between borrowers and lenders. Lenders can be commercial banks or shadow banks. Shadow banks finance the mortgage with money borrowed from other banks or investors. All mortgages are sold to one of the Wall Street investment banks, such as Lehman Brothers, who borrow money to buy a pool of mortgages. Lehman then make a profit through the cash flow arriving from borrowers.

In addition, Lehman Brothers could create MBS as bonds, like any other bonds. The MBS would be sliced into smaller bonds to be sold as collateralized debt obligations (CDO) to different investors to make a profit. These investors could then buy insurance to protect them from CDO defaults, called credit default swaps (CDS). The MBS and CDO are termed securitization. The finance transactions are handled by special-purpose entities (SPEs).

Normally, to buy and mortgage home borrowers need to have: (1) enough cash for a down payment; (2) sufficient income; and (3) other requirements. In addition, they must have a good credit score on their payment history. These are requirements for conventional and/or FHA-insured (Federal Housing Agency) mortgage loans called prime loans.

Note that not everyone can be qualified for conventional and/or FHA-insured mortgage loans. So the above criteria are relaxed for borrowers with low incomes and low credit scores. The latter are riskier and, therefore, the interest rate would be higher to compensate for the risks. This extra interest rate difference is associated with the risk of missed payments, foreclosure, and so on.

The scheme is termed subprime lending. Two additional Acts opened the doors to the subprime market: the *Depository Institutions Deregulation and Monetary Control* Act in 1980, and the *Alternative Mortgage Transaction Parity* Act in 1982. A third Act in 1986, *Tax Reform Act*, helped expand the subprime market.

To make mortgages affordable to low income and riskier borrowers, the adjustable interest rate mortgage (ARM) was introduced as an alternative to the fixed rate mortgage (FRM). Borrowers would enjoy a low introductory rate for the first few years, which would then jump to a much higher rate. Borrowers would agree to an ARM because they expected to refinance when the ARM expired. If the rate went higher, they could resell the house and make a profit, since housing prices commonly enjoyed an annual 5 % increase.

An example taken from Larry G. McDonald, author of *The Colossal Failure of Common Sense* illustrates this point. A $300,000 loan, without a down payment, at 2 % interest would pay a monthly $500 mortgage. So, with a thousand such loans Lehman Brothers could earn $500,000 monthly if the borrowers all paid their mortgages. Thus, if Lehman Brothers bought a $300M trench of subprime mortgages they could make a huge profit on one thousand loans.

From the start some subprime loans were defaulted on, but the default percentage was manageable. As it turned out, there was a rapid growth in this market in the period 1995–1999. From 2000–2004 there was a much higher volume of these riskier loans. As recorded, between 2004 and 2006 Lehman Brothers had a 56 % increase in real-estate revenues. The increase continued and, in 2007, Lehman Brothers reported revenues of $19.3B.

But from March 2007 the subprime market worsened. BNC, bought by Lehman Brothers in 2004, was closed in August 2007 due to defaulted loans. A year later, on March 13, 2008, Bear Stearns collapsed with only $2B cash on hand for the defaulted loans after lenders pulled back credit lines and investors made withdrawals.

The downturn of the subprime market began with an increase in home foreclosures and a decrease in housing prices in 2006 and started to go out of control in 2007. Interestingly, despite the market downturn, Lehman Brothers had grown from $2.6B in 1995, to $11.2B in 2000, and $51.8B before it crashed in 2008. Like Bear Stearns, Lehman Brothers was unable to keep making payments to its loans with its cash flow.

Lehman Brothers had to use a mechanism called Repo 105 (for fixed income) and Repo 108 (for derivatives) to raise cash, via SPEs. The Repo works as follows. Lehman Brothers would sell its assets to Hudson Castle (a fictitious organization) and repurchase them within a couple of days at 105 % of the asset value. Generally accepted accounting principles (GAAP) would allow the transactions to be made as sales; therefore the assets could be removed from the balance sheets. The only problem was that Hudson Castle was fictitious. It was controlled by Lehman Brothers. It was a fraud. Ernst & Young was involved but failed to report. This was similar to the Enron scheme in creating SPEs to hedge funds in 2001. Enron SPEs required 3 % from an independent investor, however, in reality the 3 % was financed by Enron employees.

How It Finally Happened?

Tracing back from the bankruptcy filing date, there were several activities during the week prior. On September 9, Lehman Brothers stock lost almost half its value after refusing an offer of $23 per share from the Korea

Development Bank (KDB). KDB failed to attract partners to buy Lehman at a higher price. On September 10, Lehman Brothers announced a loss of $3.9B and their intent to sell part of their assets. On September 13, there was a meeting called by the Federal Reserve Bank of New York in an attempt to help find buyers. On September 14, it was reported that Barclays had backed out due to an objection from the Bank of England. Another potential rescuer, Bank of America, decided not to help Lehman Brothers.

Why It Happened?

The market drove Lehman Brothers into subprime lending. Lehman Brothers wanted to play big. Lehman Brothers was very aggressive. It bought BNC in 2004 to handle subprime lending directly, but had to borrow from others to finance MBS and to sell CDO. It was short of cash. Further reasons leading to the collapse included: the government's refusal to bail it out; the Bank of America's refusal to buy; and Barclay's inability to obtain support from the Bank of England. The root causes were described by D'Arcy in 2009 and labeled as the three L's: *leverage index*, *liquidity* and *loss*.

In the Lehman Brothers' case we can identify three issues:

1. *Exception issue*: Lehman Brothers' people were top-notch in the business from a strategic to an operational level. Its VPs and managing directors were responsible for all areas of investment, trading, risk, etc., including a strong and capable research and analysis team. Lehman Brothers kept their professionals alert through weekly meetings with the best information and analysis for their traders, covering all areas of their business. From the managing directors down, people were kept on their toes and were very productive. The full acquisition of BNC in Irvine, CA, and Aurora Loan Services in Littleton, CO, had originated $40B in the subprime market. However, the housing market went south. It allowed and included mortgage loans nicknamed NINJA (no income-no job-no asset). People were able to get loans without documents, while the price of housing was rising.

   Red flags started to surface internally when Alex Kirk, a VP of Lehman Brothers, predicted bad news in the housing market. There were red flags all over during 2008 but Richard Fuld Jr. ignored them. The Bear Stearns collapse in March 2008 did not slow Fuld down. He did not listen to VPs or managing directors, other than those closest to him. Again, there were information disconnects, Fuld's abuse, and negligence in management control (in a different way to the Barings case).

2. *Decision issue*: There was the mentality that top executives could do whatever they wished in this hostile environment. Lewis Glucksman, co-CEO, ousted his mentor of 10 years abruptly and ruthlessly. It was power, expansion, and money, pure and simple.

   Nevertheless, in the early 1980s Lehman Brothers was prosperous. They expanded and spent on capital investment, operating expenses, and technology. Then the business headed south and underwent an American Express rescue for 10 years. In the mid-1990s Lehman Brothers spun off from AMEX with Richard Fuld Jr. as CEO. From then until 2008, it was thought that Richard Fuld Jr. was operating as he had done in the 1980s. He was isolated from the rest of Lehman Brothers. When short of cash in 2008, he made a series of bad decisions: authorizing Repo 105 to hide a $50B loss; refusing an offer from the Korea Development Bank; faking an account of $500M in the Bank of America. Richard Fuld Jr. sincerely believed he had buyers or the FED would have bailed it out. He was quite wrong in thinking that Lehman Brothers was too big to fail.

3. *Control issue*: Lehman Brothers was one of the four largest investment banks on Wall Street. Its people were money-driven, but the environment was hostile. The Board consisted of only ten members so their was no one to challenge Richard Fuld's decisions. There was a big issue of corporate governance.

Summary and Discussion

New laws and regulations were carefully drafted from lessons learned in response to the collapse of Enron and Lehman Brothers, such as the Sarbanes-Oxley Act of 2002 or the Dodd-Frank Act of 2010. Revised roles of executive directors, revised rules on derivatives trades and agreements were written by the largest banks, and accounting, financial and legal reforms were proposed and enacted. Some authors reported on successes of the new regulations and reforms. Others cited discrepancies among reforms, expressing doubts and reporting failures.

The Sarbanes-Oxley Act was enacted to strengthen CEO and CFO responsibility and accountability, professionalism of auditing arms, and code of ethics. Yet, 6 years later, Lehman Brothers top executives were involved in severe frauds in September 2008. Collectively the regulations and reforms have been less than satisfying. Most importantly, there is not yet a comprehensive theory of prevention, that we know of, targeting fiasco or bankruptcy prevention in an institution.

Nevertheless, as shown at the beginning of this chapter, fiascos come in two flavors: with fraud and without. We have seen multiple cases with frauds: Barings, Enron, WorldCom and Lehman Brothers. There have been others, including: Adelphia, Tyco International, Parmalat, Société Générale, Bear Stearns. Without fraud, we can name US Army Future Combat Systems, US Air Force ECSS, US Marine Corps GCSS, or Healthcare.org, in addition to others listed in Table 2.2.

Fraud or no fraud, these institutions all shared a common problem, questionable management leadership making questionable decisions, which produced questionable strategies, policies, processes, practices, operations, and so on.

New regulations and reforms would have worked to some extent. In non-fraud cases, future fiascos and bankruptcy may well be caused by bad management and bad leadership, with good people making wrong decisions,and/or honest mistakes. In intentionally fraudulent cases there is always someone who will break the law. This could be due to irrational, emotion-driven decisions. It could be due to information disconnects. In most past cases, management abuse may have been exercised for the same reasons: greed, power, risk, and so on. Corporate governance may have been limited in interfering successfully due to organizational structure and functionality.

No doubt new fiascos will occur and new reforms will be devised. Existing reforms will be fined-tuned, and the vicious circle will continue to spiral. After all, everything is driven by human decision makers at all levels of the institution, market and economy. And, being human, we are subject to a wide range of behavior driven by human mind, from extremely good to extremely bad. The human mind should be the key concern.

We need systemic problem solving. We need a systemic foundation. We identify three areas of concern: exceptions, decisions and control. We need some systemic way to see *signs and symptoms* of unusual happenings, much like the signs or symptoms of disease in the human body. The signs must be detected early. The symptoms must be felt to respond to early enough. We call these, collectively, *exceptions*. The situation must be understood.

Human *decisions* and the decision-making process must also be understood from the decision maker's perspective. We set out to look into what can help detect signs and symptoms of a potential fiasco. We would like to know what is in the mind of human decision makers. We will question how we can organize a systemic framework and a conceptual model with adequate functionality to exercise fiasco prevention based on these two main concerns: *exceptions* and *decisions*, which intertwine in what we call the *exception-decision complex*. To handle the complex in the institution, the *control* must be extended to an organizational unit different from the main line of command for checks and balances beyond the board, accounting audit, legal consulting and the like.

CHAPTER 3

Preventing Corporate Fiascos:
A Systemic Approach

In the fiascos and bankruptcies of the institutions illustrated in Chap. 2, we looked at the employees as the main driving force. They are the decision makers whose decisions move the institutions from point A to point B. We would like to understand, based on what employees are chartered to do and the decisions they make, under internal and/or external constraints. We need to understand the environment the institution is part of, i.e. the market and the associated economy. Thus, we need to consider the institution within a much larger scope than itself.

Our proposed scope is the biological spectrum, which consists of nine components in a hierarchy: protoplasm (lowest level); cell, organ; organ system; organism; population; community; ecosystem; and biosphere (highest level). This spectrum encompasses almost everything and anything that nature and humans create or do. We single out six of the nine components as of particular interest to us. These are *protoplasm* in cells, *cell* in humans, *human* as a type of organism, *institution* as a type of community, *market* as a type of ecosystem, and *economy* as a type of biosphere.

We apply the concept development by analogy and exercise the analogical reasoning method suggested by Paul Bartha (Bartha, 2013) and by Lyndley Darden (Darden, 1982) for the identification of the analogues within the spectrum. We justify our analogies with facts taken from the cases discussed in Chap. 2. Hereafter, when referring to a particular component, such as a *cell*, we use two terms interchangeably, *cell* or *cell-human*. The compound term, *cell-human*, indicates that the cells make up

© The Editor(s) (if applicable) and The Author(s) 2016
T.N. Nguyen, *Preventing Corporate Fiascos*,
DOI 10.1057/978-1-137-49250-0_3

the *human* component, which can be viewed either from a cell perspective (bottom up) or a human perspective (top down).

The main purpose of this chapter is to establish an empirical basis for a theory of prevention. This includes: (1) a systemic framework derived from the biological spectrum; (2) a conceptual model for the prevention of fiascos in institutions, on which systemic thinking can be exercised; and (3) organizational control for prevention.

ON A BASIS OF FIASCO PREVENTION

The Foundation: Figure 3.1 Explained

The box on the left of Fig. 3.1 depicts the nine-level skeleton of the von Bertalanffy-Boulding general systems theory (GST) (von Bertalanfyy, 1950; Boulding, 1956) for the unity of science. GST has been an inspiration for many researchers in developing systems thinking applications in numerous disciplines.

Von Bertalanffy's research identified the existence of isomorphic laws, i.e. those laws common to all disciplines (physical sciences, social science,

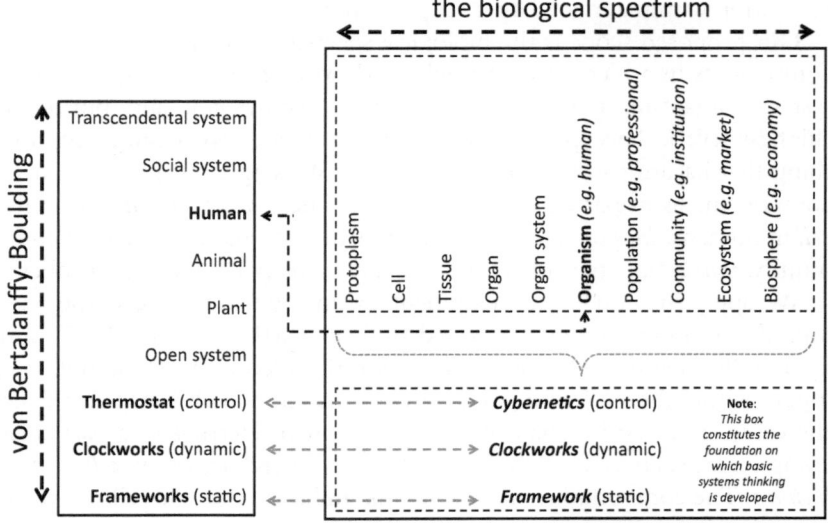

Fig. 3.1   The biological spectrum foundation

philosophy, etc.). An example of an isomorphic law is the exponential law describing the decay of radium (microscopic world), or the dynamics in population (society), or in compound interest (finance). From these isomorphic laws, he postulated general system laws applicable to systems of any type.

He defined a system interacting with its environment as an *open system*. He then formulated important concepts based on principles he found that were similar and valid for all systems. One concept being *equifinality*, which expresses that the final state of an open system may be reached from different initial conditions via different ways. The opposing concept is *multifinality*, which means that initial conditions might lead unexpectedly to different results. From our perspective, one such unexpected and different result could be a fiasco.

In Boulding's version, Kenneth Boulding used the term *frameworks* (in plural) and the corresponding *clockworks* to indicate respectively the static and dynamic aspects of open systems. A particular framework, according to Boulding, can be microscopic or macroscopic, as can its corresponding clockwork. *Thermostat* is Boulding's reference to the cybernetics of Norbert Wiener equivalently (Wiener, 1948). The objective of a thermostat is to work with the clockwork in maintaining a set of equilibrium states as the limiting case of the dynamics in clockworks. All three (frameworks, clockworks on the frameworks, and thermostat on the clockworks) constitute the base levels for living species. Life begins at the *cell* level (considered in Boulding as the *open system* level), with its self-maintenance and self-reproduction.

The next two levels (plant and animal) are self-explanatory and are not considered relevant to our formulation, therefore are skipped in our discussion. The level immediately above animal, *human level*, is identified by Boulding with a focus on the human ability to think and with thoughts expressed using language and speech for communication. The social, or social organization, level which humans make up, according to Boulding, is more about the role of the humans than themselves. The roles are interrelated via the communication of messages, whose contents and meaning are of concern. The highest level, the transcendental system, is about the unknowables. We discuss the formulation of the last four levels differently in the context of systems thinking for decision making in our proposed framework and model.

In our approach, first we restructure the combined von Bertalanffy-Boulding's skeleton of science as shown in Fig. 3.1. The box on the right depicts our overall scope. Its top black dotted box is the biological spectrum presented earlier. In this way, we do not consider the transcendental system. The social systems in von Bertalanffy are replaced by four components: population, community, ecosystem and biosphere. The biosphere is the highest level component considered here.

We consider one big static framework (singular) in the bottom right dotted black box as equivalent to von Bertalanffy-Boulding's different frameworks. There are multiple clockworks (dynamic) within this unique framework however. Each clockwork exists and corresponds to its associated component using a modified version of cybernetics originated from Norbert Wiener.

There are many versions of cybernetics which have extended Wiener's original concept. Ross Ashby (Ashby, 1957) developed molecular cybernetics at cell level, an adaptation of Wiener's cybernetics for cell metabolism. Stafford Beer, in 1972 (Beer, 1972), worked on managerial cybernetics at institution level. Robert Hoffman in 2010 Hoffman, 2010) expanded it to economic cybernetics at an economic level.

We scope the three domains—*framework* (static, singular, overall), *clockworks* (dynamic, component-focused) and *cybernetics* (control, communication, feedback, component focused)—as the foundation (bottom right black dotted box) on which: (1) all other components of the biological spectrum are based; and (2) within which some basic systems thinking is developed (to be listed in Fig. 3.4).

In the *biological spectrum*, sketched in Fig. 3.1, we observe that each component involves a number of major disciplines. The protoplasm-cell component is the subject of molecular biology, medicine, pharmacy, and so on. The cell-human component involves biology, chemistry, physics, and so on. The upper three components (human-institution, institution-market, and market-economy), represent such disciplines as business, law, political science, and economics, to name a few. The human-institution component focuses on organization, management, psychology, and related sub-disciplines. Institution-market involves marketing, accounting, finance, law, ecology. Market-economy includes political science, economics, and other disciplines. A higher component in the network subsumes the ones below it.

Each component has been well developed in its own right with theories, methods and techniques for the related disciplines. There have been attempts to: (1) learn a property or process from one component to apply it to another; or (2) integrate various components in some fashion. In fact, in the old days, Von Neumann and Burks applied cellular automata to computer science. Maturana and Valera used biological concepts to develop autopoiesis theory, which was a way to explore emergence property (Maturama, 1980). James Moore used ecology to build his concept of business ecosystems, which resulted in the European Consortium studying a new breed of systems called Digital Business Ecosystems (Moore, 1993, 1996, 1998). Recently, economists turned to neuroscience to link economic decisions to corresponding brain activities in decision neuroeconomics as described in Paul Glimcher (Glimcher, & Fehr, 2013).

It can be expected that upper (higher-level) component will be affected by lower components, structurally, functionally, and behaviorally (bottom-up). It implies that a change in a lower component can yield either (1) no effect, or (2) affect any or all of the higher components. The effect can be positive or negative, direct or indirect. It is similar to the case where a virus hosted by a cell can lead to a disease affecting a human organ, organ system, or the entire body. Another case is where an abnormal cell whose genes are mutated externally or internally can invade nearby tissues to become a malignant tumor. The latter might grow uncontrollably in a process called metastasis, this might lead to cancer which proliferates to one or more organs.

By the same token, a human employee whose thinking is psychologically different for some reason, can lead to human decisions which affect the institution's behavior. This is similar to the case of Nicholas Leeson who decided to hide big losses in an error account.

It is also the case of Enron executives who decided to use SPEs to transfer losses and fake gains in their quarterly and yearly financial statements. By committing fraud they eventually caused Enron's bankruptcy. Besides the accounting fraud, Enron executives also manipulated California's electricity supply, creating an energy crisis which eventually led to the collapse of the Pacific Gas and Electric Company.

Bernard Ebbers and Scott Sullivan also caused a WorldCom fiasco through fraud. They initiated turmoil in the telecommunications market in which WorldCom is a major player.

Richard Fuld Jr. and his team, who exercised Repo 105 and lied about the funding of Bank of America, brought chaos to the economy. All these cases exercised severe fraud. The effect is what Rick Looijn Looijen, 2009) identified as *complementarity* (bottom up) property in which lower components (institution) affect higher components (market and economy). The effect can ripple via networked links among institutions in the market and economy, as perceived by Albert-Laszlo Barabasi (Barabasi et al. 2004).

In summary, the basis of our approach includes the restructuring of von Bertalanffy-Boulding to observe the biological spectrum as the only static framework. Each component in the biological spectrum framework then observes its own clockwork and specific cybernetics. Together they aim to maintain the corresponding equilibrium in the component. Inadvertently, the human component can turn the institution component into a fiasco, which then affects other higher level components.

An Analogy-Based Characterization for Prevention

People frequently use metaphors to explain or differentiate concepts. Someone has explained the difference between disease and illness as follows: "*When you go to the doctor, you have an illness. When you are out of the doctor's office, you have a disease.*" An explanation of the difference between economic recession and depression is, "*If your neighbors lose their job, it's a recession. If you lose your job it's a depression.*" Sometimes these metaphors are not obvious, although they seem to trigger some thoughts.

People also use analogy and analogical reasoning to explain difficult concepts. Charles Darwin used "*an analogy between artificial and natural selection to argue the plausibility of the latter.*"

In this section, by investigating similarities and differences between all components of the biological spectrum, both in breadth and depth, we attempt to characterize our proposed basis. The sketch in Fig. 3.2 shows the analogy-in-depth (more details within a component) among three components (analogy-in-breadth, details across the three components). It requires some explanation.

The analogy of the three guiding principles between the components (human, institution, and market) is listed in the top box (indicating top layer). In the human component on the left of the top box, the first principle is the *milieu interieur,* or internal environment, a concept of the human body by Claude Bernard (Gross, 1996). It is drawn from the fact that all cells of the human body bathe in an interstitial fluid. The *cybernetics* concept authored by Norbert Wiener expresses the control and communication with feedback in the cell metabolism (Wiener, 1948). We include the concept of *homeostasis* defined by Walter Cannon as the goal to be maintained in a healthy human body (Cannon, 1963).

	Human		Institution		Market
Guiding principles (*top view*)	Body fluid, Plasma ←→ → "milieu interieur" (Claude Bernard)		Data environment → information management	←→	*Information environment* → *financial management*
	Cybernetics (Norbert Weiner)	←→	Managerial cybernetics (Stafford Beer)	←→	*Economic cybernetics* (Robert Hoffman)
	Homeostasis (Walter Cannon)	←→	Stability	←→	*Equilibrium*

⇕

	Human		Institution		Market
Organization (*mid-view*)					
Structural	Cells, Tissues, Organs	←→	Employees, Professionals, Departments	←→	*Institutions, Populations*
Functional	Organ systems	←→	Divisions	←→	*Communities*
Behavior	Biological processes	←→	Business processes	←→	*Economic processes*

⇕

	Human		Institution		Market
Supporting entities (*bottom view*)	**Cells**		**Employees**		*Institutions*
	Macromolecules	←→	Projects	←→	*Contracts*
	Interstitial fluid	←→	Data	←→	*Information*
	Cellular exchanges	←→	Transactions	←→	*Corporate transactions*
	Blood	←→	Budgeting	←→	*Hedging*
	DNA (genes)	←→	Policy (regulations, rule)	←→	*Law (regulation, SEC*)*

* SEC: Securities and Exchange Commission

Fig. 3.2 Analogy between human, institution, and market

At the institution and market component levels, we identify the *data environment* in the institution where human employees work on a daily basis and the *information environment* in the market on the right of the top box. Both are respectively analogous to the *milieu interieur* in the human component. Similarly the analogues to Wiener's cybernetics in the human component are *managerial cybernetics* and *economic cybernetics*. The analogues to homeostasis in humans are *stability* in the institution and *equilibrium* in the market. Those mentioned principles constitute the analogous guiding principles at the top (or strategic) layer of the three components: human, institution and market (Figure 3.2).

Similarities for analogies in the organization layer (structural, functional, and behavioral) in the middle box of Fig. 3.2 are self-explanatory. The supporting entity layer in the corresponding clockworks of the three components is shown in the bottom box of Fig. 3.2.

The entities (macromolecules, interstitial fluid, cellular exchange, blood, and DNA/genes) at the *cell-human* component are analogous to the entities (project, data, transaction, funding, and institution policy) respectively in the *human-institution* component. They are also analogous to the entities (contract, information, corporate transaction, hedging, and government laws and regulations) in the *institution-market* component. The following explains the analogues briefly.

At the cell-human level, the proteins, a type of macromolecule, in cells account for more than 50 % of the cells' dry mass. Others are carbohydrates and lipids for storage, and nucleic acids as codes. Proteins of the macromolecules in the cell-human component are constructed from 20 different amino acids.

The proteins in the macromolecules are similar to activities or projects in the human-institution component. The cellular exchange therefore is similar to transactions between employees responsible for a project in the institution. The nucleus in a cell is like the brain in a human. The organelles working for the health or well-being of the cells are analogous to the processes employees perform for the well-being of an institution. To depict the analogues of blood and DNA (genes) in cell-human component we use funding and policy in the human-institution component and hedging and regulations in the institution-market component.

The structural analogy at three different views or layers (guiding principles, organization, and supporting entities) among the three components (human, institution, market as listed in Fig. 3.2) gives rise to the application of processes known in one component to another component, when applicable. We could say that fiasco development in an institution could be considered as analogous to the development of cancer in humans, which we explain further in the next subsection.

In brief, we explore topological isomorphism (via analogy), rather than von Bertalanffy's isomorphism; we restructure the components in terms of the biological spectrum, rather than Boulding's skeleton of science. We have identified guiding principles, organization, and supporting entities in each component analogous to other components with a focus on human, institution, and market. Also, different to von Bertalanffy-Boulding's framework in which all levels are considered equally important, ours considers the human component as key, one in which and from which fiascos are initiated. One could say that our systemic approach is human-driven.

Cancer Analogy

This analogy sets the stage for an overall framework for exploring the fiasco process in an institution. We now see, via analogy and analogous reasoning, how our framework (static), with clockworks (dynamic), and cybernetics (control and communication), behaves in an abnormal environment, in terms of cancer in humans and fiasco in institutions.

First, we define as *norms*: (i) *normality* (not labeled in Fig. 3.2) in protoplasm; (ii) *constancy* in *cells* (not shown); (iii) *homeostasis* in the *human body*; (iv) *stability* in *institutions*; (v) *equilibrium* in *markets*; and (vi) *balance* in the *economy* (not shown). Any deviations from the norms at any component level are signs and/or symptoms of *exceptions* in the component level and all levels above it. By exceptions, we mean any faulty events which occur.

The structural analogy in Fig. 3.2 does not tell us much in terms of deviations from the norms. We need to look at the same analogy from a different perspective, that of functional and behavioral in terms of activities, exceptions, and control. An example is drawn from the cell-human component where we reformulate its clockwork environment, as shown in Fig. 3.3a. Deviations from the norm are captured as exceptions.

The identified supporting entities in the cell component, namely macromolecules, blood, cellular exchange, and interstitial fluid, are all governed by DNA and gene expressions. The control exercised by the human brain and mind are measured against the norm. The norm in the human component is the *homeostasis* to maintain the health and well-being of the human (Fig. 3.3a).

A human body within the biological spectrum can deviate from its homeostasis as a result of anomalies in its anatomy, physiology, or diseases contracted. Some diseases are deadly. One particular deadly disease in humans is cancer.

A normal cell can turn cancerous for external (e.g., radiation) or internal (e.g., gene mutation) reasons. When cancerous cells invade nearby tissue, they are classified as a malignant tumor. The latter can quietly proliferate to other organs via blood or lymphatic circulations. When cancer symptoms are detected the cancer is, commonly, already in a more critical phase, which eventually can cause death.

If an institution (business, industry, or government) is thought of as a human body (Fig. 3.3b), its employees can be considered as analogous to human cells. The corresponding support entities at the human-institution component from the activity (or its resulting event) perspective are the projects executed according to a business process, data environment, funding, or transactions reported in corporate financial statements. The norm considered at the institution component level is *stability*. Any deviations from the norm (events or decisions) with respect to the control gov-

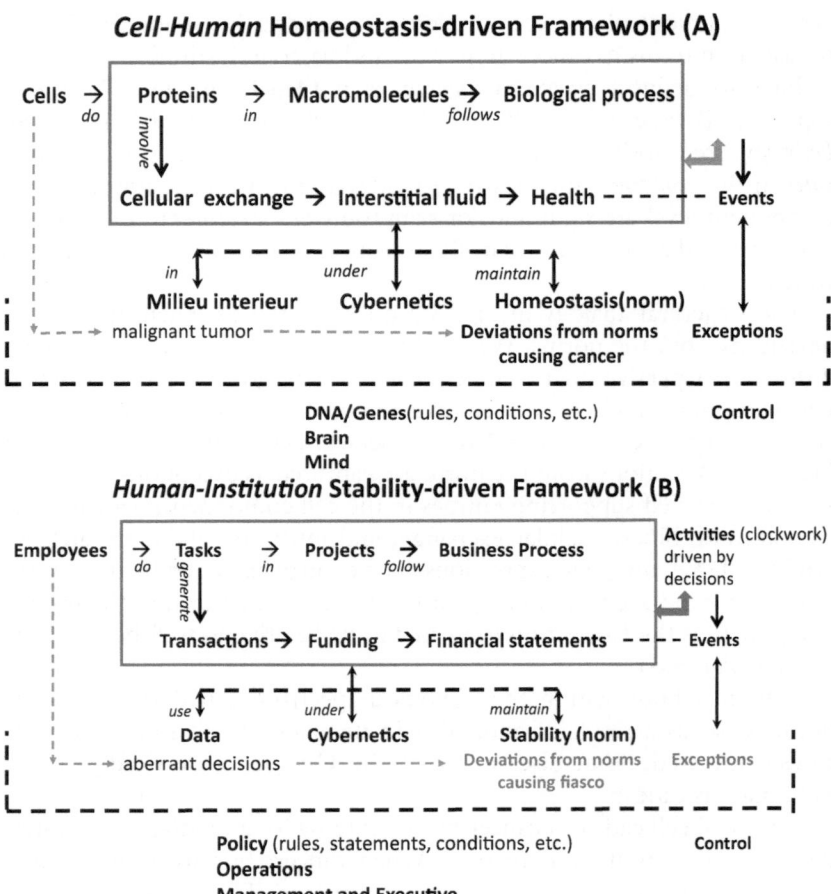

Fig. 3.3 (a) and (b): Analogy between cell-human and human-institution models

erned by corporate policy, or exercised by management and executives, are considered as exceptions.

Any employee in the institution can turn *abnormal* for any reason, much like any cell in the human body. Leeson of Barings Bank was considered abnormal as he was driven by self-interest. He wanted to be a successful trader. Trading losses were unacceptable to him so he abused the use of error account 88888 to hide the first loss of £20K caused by Kim Wong (Leeson, 1996, 2012).

For other people, the reasons could be incentives, greed, business culture, power, and so on. Eisenbardt discussed incentives as the major driving force of employees in his agency theory. Jeff Skilling promoted incentives in Enron's personnel review system. Other authors recognized greed as a driving force in the Enron case. Andrew Fastow created a series of special purpose entities (SPEs) designed to hedge funding in support of the expansion of the gas bank concept to other industries (CLayton et al. 2002).

The WorldCom case was driven primarily by a business culture created by CEO Bernie Ebbers. On the other hand, it seems fair to label CEO Richard Fuld of Lehman Brothers as geared towards power. They all were driven to commit an initial deviation from the established norm in the corporate policy or standard operating procedures.

If the deviations and their consequences are ignored, tolerated, misunderstood, misguided, or involved in some further act of covering up, they can lead to more severe situations. In the Barings case, Leeson continued to hide the losses in the error account without anybody's knowledge, except the loyal, local employees who worked for him and who had no accounting background.

If employees who behave abnormally or erroneously, as illustrated above, also have managerial authority they can exercise their influence beyond the scope of their organization unit, which is analogous to an invasion. Leeson had a dual role, general manager and authorized trader, therefore he was able to maliciously alter the financial losses.

If their influence and the bad outcomes proliferate to other organization units—including regulatory, legal, financial, accounting, managerial—in corporate governance, they commit a severe fraud. This is what Andrew Fastow did to Arthur Andersen and his legal team.

By the time the signs or symptoms of fraud are detected, much like the cancerous symptoms, it might be too late. The institution will collapse. This was almost exactly what happened in the Barings and Enron cases from a cancer and fiasco perspective.

The collapse of an institution affects the market which is driven or supported by that institution. This market can involve various hierarchies of structures, as seen by R. Coase, which the institution is a part of (partnership, business alliances, etc.), or an ecological complex, which James Moore called a business ecosystem. An abnormal employee can cause a market or business ecosystem to be *out of equilibrium*, which would affect the economy.

Additional Insights from Cancer Analogy into Fiasco Prevention

Cancer in humans has the following features: (a) it fosters uncontrollable growth; (b) human lifestyle is a major influence; (c) it is about change; (d) it can invade and proliferate to other sites; and (e) despite the fact that it is extremely difficult to cure, it is not always lethal (i.e., it can be fixed if known about early enough). These features of cancer offer insights into how we could handle corporate fiasco.

Uncontrollable Growth, Invasion, and Metastasis

Cancer cells are normal cells whose genes are mutated or damaged, for external or internal reasons. They start to produce other monoclonal cells by division or mitosis, which, subject to additional mutations, can show uncontrollable growth.

The financial growth of Barings Singapore occurred in a direction more negative than positive. The handling of Philippe Bonnefoy's account and the arbitrage activities by other traders had positively brought in profits. On the other hand, the cover-up of Kim Wong's and Goerge Sowe's losses of £70K dragged into more losses during 1993 and part of 1994.

Although Leeson was able to recover completely by balancing out error account 88888 to zero in July 1994, he had to resume his covering-up activities during a bad trade on behalf of Philippe Bonnefoy. The early loss amounted to £50K. Nicholas Leeson then became uncontrollable. When he was on the verge of losing Philippe Bonnefoy's account, he took the risk of offering counter calls at .139 or higher, rather than at .138, and the market did not favor his take. He was unstoppable in trading Nikkei Index 225 in Osaka and SIMEX. Leeson made big losses.

A similar story of uncontrollable growth applies to Enron in the handling of SPEs and to Lehman Brothers with Repo 105. Fastow influenced practically everybody in his team (analogous to tumor invasion) in extending the fraud. He also influenced the accounting firm of Arthur Andersen and the Vinson and Elkins legal team to work in his favor (analogous to metastasis). In Enron, *invasion* and *metastasis* were not detected until October 2001, in its third quarter report.

Life Style and Risks

Cancer is greatly influenced by the human life style. The American Cancer Society reported a link between cancer and alcohol, cigarette smoking, or exposure to UV in the sun. There are several other risk factors for cancer, such as asbestos, radiation, and the like.

Similarly, there are risks of fiascos in institutions driven by human-employee life styles. These can be observed in the behavior of Barings employees in their trading activities or of Enron executives in MTM activities. These risks driven by an employee component can be bottom-up (from employees at any level to institution, similar to gene mutation to cell), or can be top-down, as in the risks of cancer which can be caused by human functionality and behavior (top-down, from human behavior in life style down to its constituent cells).

Cancer, Symptoms and the Immune System

Cancer and its symptoms are hard to detect because they occur below the level of human consciousness and detectable only by the autonomic nervous system.

One could ask: *"How about the human body's immune system? Could it detect and destroy cancer cells?"* The immune system consists of a collection of different organs, special cells, and membrane substances working together to protect the body from disease, infection or invasion. It works well, but when it comes to cancer cells it does not.

In fact, the immune system has the ability to recognize strange substances and raise alerts. Cancer cells, like germs and other kinds, have a special substance in the form of proteins, but unlike the others, cancer cells cannot be easily detected by the immune system. This is because the system does not recognize that the special substances in cancer cells are foreign invaders. To the immune system they are more like traitors from within. Therefore they go undetected.

By the same token, nobody in Barings bank suspected that Leeson would commit fraud. He was a collaborator. He had been among the top performers since his first assignment to Indonesia settlements. He had cleaned up the messy back office there. Baring executives and management trusted him, and never suspected he was up to any fraud. To detect a fiasco we need to add a role analogous to the immune system, with enhanced functionality to identify those beyond *traitors*.

To address the issue of un-detectability there are drugs that mark cancerous cells. Thus, one possible way in the institution component is to mark exceptions, decisions, or the exception-decision complex (Fig. 3.3b) with a criticality attribute. The values of the attribute can be marked as *warning* or as *severe* (qualitatively), and/or accompanied by a numerical value, (quantitatively). This is analogous to the antigens on the surface of a cancerous cell.

Systems Thinking (Descriptive) on Clockworks: Figure 3.4 Explained

Figure 3.3a, b shows the clockworks of two components: cell-human and human-institution. The institution-market and market-economy components can be similarly formulated. We now look into how we can exercise systems thinking on each, and among the clockworks, in the overall framework.

The right box labeled *systems thinking* in Fig. 3.4 is our overall approach to problem solving. It consists of component system thinking slightly different one from the other due to the specificity associated with each component.

Depending upon where the problem resides (in cell, human, institution, market or economy component), the systems thinking has an additional focus, and is labeled as such in the smaller box on the right of Fig. 3.4. At the cell component we can talk about cellular systems thinking—consisting of the process of protein synthesis, the Kreb cycle for all metabolites, and so on—as directed by the cellular nucleus—the brain of the cell.

From the human component up, we explore the human abilities recognized by Kenneth Boulding in his skeleton of science mentioned previ-

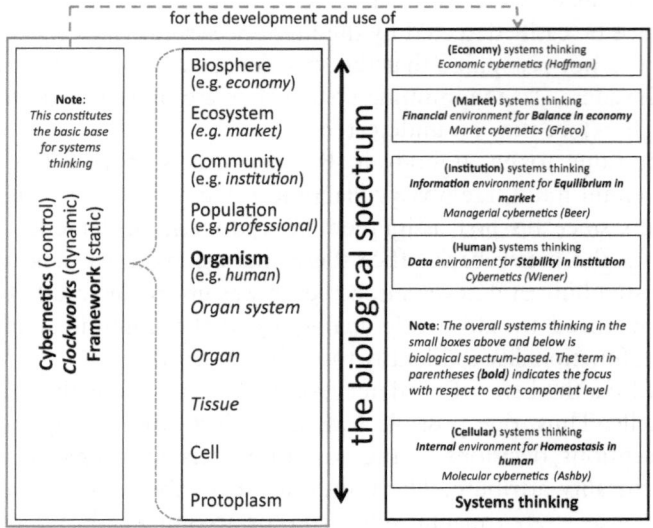

Fig. 3.4 The biological spectrum and systems thinking

ously. Specifically the one called decision making. It is this crucial ability which also drives the expected equifinality in its component and in the upper components of the biological spectrum (market and economy). We will discuss the decision making process driving all human decisions further.

The analogy of Claude Bernard's internal environment to *data* (employee concerns), *information* (institutional concerns), and *finance* (market concerns) in support of the economy component level, is derived from an analogy with the interstitial fluid within the human body. While the human body needs oxygen to reach the lungs, to oxygenate blood in the heart, to carry it to all cells to keep them alive, the institution needs information to make decisions, to generate money, to keep all employees alive.

The second guiding principle on control, communication, and feedback in Wiener's cybernetics was extended by Stafford Beer in 1972 and Robert Hoffman in 2010, which we will explain further.

While Beer extended Wiener's cybernetics to the institution component level, Hoffman discussed mainstream economics and cited some reasons for failures. Hoffman proposed a reformulation of economics as a management of entities called *commons*. The focus on management makes the economic problem a cybernetic concern. Hoffman's well-defined proposed objective involves the concept of balance as the norm we injected in the framework. A continuous observation of the state of commons is analogous to our proposed exceptions management. Hoffman's management of markets in an economy with an understanding of decisions made on a continuing basis would fall into our need to understand corporate decisions. James Moore's concept of business ecosystems in the economy also falls into our understanding of impact and the ripple effect of corporate fiasco on the market and economy.

The descriptive systems thinking sketched out in Fig. 3.4 allows us to address logical systems thinking on fiascos at the institution component or higher by borrowing known processes in lower components.

A CONCEPTUAL MODEL FOR PREVENTION

As discussed, cancer in humans is a deadly disease and difficult to cure. We should not wait until it is too late. So the issue here is not the management or cure of cancer but its prevention. The conceptual model is based on four interrelated concepts, as detailed in the following: *overall framework, exception management, understanding decision,* and *oversight organization unit.*

Conceptual Modeling: A Framework and Model for Prevention

In search of the fiasco prevention process we can start to center on an understanding of the analogues between components. One way is to detail the structure, functionality, and behavior of each component in the spectrum in its normal working environment in terms of constancy, normality, homeostasis, stability, equilibrium, and balance. It involves an attempt to understand how each lower component interacts with other components in the spectrum: sideways, upwards, or downwards.

We note that if this approach is taken the complexity will increase enormously. The reason being that a higher level component is influenced not only by the one immediately below it but also by all components at lower levels, directly or indirectly, as expressed in the property of *complementarity* in nature—a daunting task.

Therefore, we take a different approach to the problem. As alluded to in the previous section, we focus on the *abnormal* behavior of entities in the institution—called *exceptions* as shown in Fig. 3.3b—rather than on their normal behavior. This, in essence, reduces the size and complexity of the problem domain of interest.

We propose a conceptual model for fiasco prevention in an institution that includes three types of entity. The first type is *exceptions* to be detected and made transparent as they occur, subject to validation, analogous to symptoms of cancer or disease. To be detectable, we add a function similar to that of the autonomic system in the human component for the detection and reporting of exceptions in the institution component. This is an addition to any integrated information management by exceptions which might already exist in the insitution.

Second, *decisions* which cause the exceptions, or fix them questionably, are to be evaluated for an understanding of their root causes, preferably from the decision maker's perspective. This approach is new in the domain of corporate decision making.

Third, we propose an *enhanced* management control. We cannot leave control in the hands of executives and top management since they might exercise abuse, greed, power, and so on, as we have seen and discussed previously in Chap. 2. To this end, we add an *Oversight organization unit* charged by the Board of Directors. This unit operates independently but alongside the institutional organization structure. This addition is also new in corporate governance.

The three domains in the framework—exception management, understanding decisions, and an Overisght organization unit—together with the

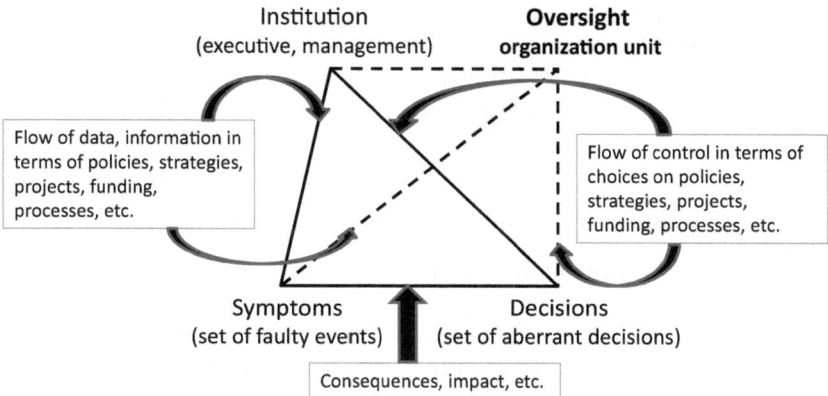

Fig. 3.5 Conceptual model for prevention

institution are illustrated by Fig. 3.5. This conceptual model is intended primarily to foster a prevention scheme analogous to cancer prevention. We explain the three domains below.

Conceptual Modeling: Exception Management

Our approach suggests at least three issues to be considered by the exception management: we need to detect exceptions, which is a *monitoring issue*; then we need to expose exceptions, as they occur, to all responsible parties, which is *a transparency issue*; and lastly, we need to know whether the exceptions are valid, which is a *validation issue*. Commonly, all three issues (*monitoring, transparency* and *validation*) can be handled by an enterprise risk management (ERM) in conjunction with tools such as a management by exception (MBE) system. We suggest an extension and full integration of existing corporate information systems for exception management.

The entities of interest to exceptions management are primarily, but not limited to, violations of any guiding principles, organization, or supporting entities as elaborated previously (in Fig. 3.3), such as projects, data environment, transactions, funding, and policy. An example follows. In Enron, a partnership, Chewco Investment Limited Partnership, was created to help a project called Joint Energy Development Investment Limited (JEDI II). Enron had intended to buy out the interests of CALPERS (California Public Employee's Retirement Systems) in a prior investment joint venture called JEDI I. The arrangement for the SPEs supporting JEDI II with 3 % funding was supposedly from an independent source. In reality, the investment source was not from an outsider and the managing director was an employee of Enron. These violations should have been raised as exceptions.

Conceptual Modeling: Understanding Decisions

The second part of the model involves the set of corporate decisions over space and time. The decisions might cause exceptions or they might be intended to fix the problems caused by the exceptions. This set of decisions, called D, and the set of all possible exceptions, called E, can be very complex at the institution component level since each set can be from a hierarchical or networked structure, or any combination thereof.

A logical way to handle these sets is to consider each power set, also called the σ-algebra on the set, and associate it with a numeric *measure* μ, called criticality. The σ-algebra on the set of decisions D would have two operations, *union* of decisions and *intersection* of decisions, which is closed under them. It means that the decision to be found will belong to the smaller σ-algebra on the subset of the critical decisions, the *measurable* decisions space $\{D, D \text{ and } \mu_D\}$ as well as the *measurable* exceptions space $\{E, E \text{ and } \mu_E\}$. In this newly defined space of decisions and exceptions, some fundamental probability properties, formulated rigorously by Kolmogorov, can be relaxed since our measure of criticality is subjective (human-assigned subjectively).

We need to figure out the criticality associated with exceptions and with decisions which could be subjectively or emotionally driven by the decision makers. In fact, as we elaborate on later, many aberrant corporate decisions are most likely caused by human emotion. So we explore decisions from the psychological perspective of George Kelly using the repertory grid (RG) technique, based on the personal construct theory (PCT).

The psychological approach to the elicitation, analysis, understanding and evaluation of decisions, and decision making process by the decision makers is adapted from George Kelly's PCT with some modifications. The goal is to find a criticality measure to be associated with the elements of the smallest σ-algebra of critical decisions and of severe exceptions.

PCT is intriguing to us because of its first fundamental postulate: "A *person's* *processes* are *psychologically channeled* by the *ways* in which he *anticipates events*." Kelly defined each term in *italics* in the above statement, which we can interpret to mean that the decision maker is also a process. The person's *process* in Kelly would be the decision maker's mental process in deliberating the decision alternatives.

The decision alternatives are the *anticipated events,* one of which, if selected, will follow. In Kelly, the anticipated events have a larger scope. Kelly did not restrict *psychologically* to be just psychological but it could be physiologically, sociologically, or otherwise. The channels which Kelly explained as a *"means to an end"* are important. This is why some decision makers commit a potential fraud as they believe the *end justifies the means.* Kelly's fundamental postulate is perfectly suited to our biological spectrum framework as sketched in Fig. 3.4.

Kelly's construction corollary, *"A person anticipates events by construing their replications,"* is also suitable for our proposed model. It is based on prior experience and lines of thought of the decision makers, as elaborated partially in Campbell, Whitehead & Finkelstein, and referred to as neural patterns. Kelly's construction is more powerful. He used the RG as a way to understand his patients in a clinical environment. The RG has been used and further adapted to different disciplines.

We include an attempt to locate the factors, called constructs by Kelly, which drive the decision. Via the RG technique adapted from George Kelly's PCT by others such as Valerie Stewart and used in business management, we further modify it in support of the *understanding* of decisions. The key point is the discovery of the decision maker's model of the world.

First, we need to modify the technique to address our target decision makers. Second, and most importantly, we must recognize the possibility that it might be difficult for decision makers to reveal their model of the world, especially when decision makers are unethical.

In fact, it is known that responsible parties do not always act properly on exceptions as expected. At times, they can consciously make arbitrary decisions. They can intentionally ignore, avoid, alter, or hide the detected or reported exceptions, which could aggravate or lead to fiascos and bankruptcy. Therefore it is necessary, but not sufficient, that exceptions management exposes exceptions and makes them transparent. There must be some way to "force" the decision makers to expose their thought processes for the detection of distorting attachment and misleading memories on their part. To address this issue, we suggest the Oversight organization unit below.

Organizational Modeling: An Oversight Organizational Unit

Commonly, institutions do not have a permanent Oversight organization unit. When a crisis happens, the authority assigns a one-time, ad-hoc team for crisis management. It is an "after" scheme. To prevent fiascos, we suggest an organizational unit overseeing the institution's exceptions and decisions.

The primary responsibility of this Oversight organization unit is to monitor exceptions raised and question those decisions which led to the exceptions or were made based on the exceptions. This unit is tasked by the highest authority in the institution (such as a Board of Directors). It is given the authority to demand transparency and validation in exceptions. The unit can ask for explanations of decisions made by responsible parties in the institution. This organizational unit will have its RG experts. Once exceptions are identified and decisions understood and evaluated by the Oversight unit, remedial decisions can be requested to maintain stability.

The *intertwined exception-decision* deserves some explanation. The exception-decision data elicitation, or data acquisition part, would show the what (what happened), where, when, and/or by whom. The whys can only be understood partially because of involuntary information. We need to examine the reported exceptions and their associated decisions more deeply, beyond the supporting documentation on policy, project, data environment, transaction, accounts in funding, and others—such as financial statements, negotiations, artifacts, etc. The remaining hows and whys would be revealed by the RG if properly executed, by the decision makers themselves and/or by the Oversight organization unit.

Implementation Considerations and Implication

Exceptions

The following is an example list of potential exceptions and questionable decisions to look for (the list is suggestive and incomplete). They should be substantiated by findings (documents and artifacts). The findings are similar to the way Enron information has been obtained from congressional hearings, court documents, professional associations (accounting, law, finance, etc.), with respect to corporate governance, Sarbanes-Oxley and other bills, and so on.

- Anything which is too much, too fast, too little, or too long is a candidate for further investigation: (1) Enron revenues increased at the rate of 65 % yearly from 13.3B in 1996 to 100.8B in 2000; (2) too high incentives for executives
- Any claims not supported by proper measures or indicators are candidates: e.g., Enron claimed success and profitability while accruals were negative; or free cash flow indicators are positive
- Financial analysts questioning the corporate performance are candidates: e.g., 2001 analyst report by Lou Gagliardim and John Parry of John S. Herold, Inc.; and Bethany McLean report in Fortune, 2000
- External or internal warnings are candidates: Toni Mack of Forbes in 1993; Sherron Watkins' letter to Ken Lay in 2001, etc.
- Be proactive and alert

Exception management is elaborated on in Chap. 4.

Decisions

Decision makers (DMs) have an internal, mental map or model of the world in terms of their overall experience or in terms of a specific issue (exceptions or decisions, success or failure, high risk or low risk, etc.). The DMs might use their personal model rationally or consciously based on decision analyses, quantitative methods, and other means. At times they irrationally or emotionally decide what, when, with whom, and how to proceed, while prioritizing their daily tasks. These are often indicated by "I feel that ..., I believe that ..." in their decision statements.

We have mentioned the RG technique several times as one that we can use for understanding exceptions and DM's decisions. This technique was made popular and practical for management discipline by Valerie Stewart. We should mention that although the technique is easy to understand for different applications—such as motivation, relationship, etc.—the skills for questioning DMs are difficult to acquire, and the process for eliciting the proper information is complex, especially when DMs do not cooperate. We show only some significant ideas pertaining to the understanding of DM's decisions and decision making, based on George Kelly's PCT.

The first concern is elicitation. The DMs in question are to identify a list of elements (people, objects, activities, or events) which they think are involved in a particular exception: *people* can be anybody involved in the exception; *objects, activities,* or *events* can be any projects, data, transactions, funding, policy, and so on, which exist, are performed, or have occurred.

The DMs have to select three elements at a time, called a triad. Following Kelly, we would ask:

- In what way are two elements of the triad similar (to identify *emergent or similar construct*) and
- In what way do both of them differ from the third (to identify *implicit construct*, opposite to emerging)

Towards the top of the hierarchy are those constructs which are more abstract. An example is the powerful mark-to-market (MTM) accounting concept which Jeff Skilling extended to gas bank. At the bottom are the indecomposable constructs which contribute to the exceptions, such as incentive amounts paid to executives. In between there might be something such as the constructs involved in the successful or failed SPEs used by Enron's Andrew Fastow.

A list of *emergent constructs* and their *opposites* are identified for each triad of elements during interview with the DMs. In tabular form, the columns represent the elements, and the rows the constructs. The names of the two similar-opposite constructs are placed at each end of each row. As such, the elements-constructs form a matrix or grid.

To further investigate the whys and hows of exceptions and decisions, additional ways are applicable. One is called Hinkle's *laddering up*. The question to ask is *why* the DM thinks the two elements are similar (or *why* both of them are different to the third) in terms of the identified construct. We keep asking *why* until the DM runs out of reasons. In this way, the most abstract constructs are revealed.

The other technique is Landfield's *laddering down*, which involves the *how* questions. We should ask what makes the DM think the two constructs are similar (or different). We could then relate it to other relevant constructs. The combination of the two is elaborated in *consistent laddering* by Korenini in 2012.

When the grid is complete, DMs will then be asked to rate all the constructs on a 5-point (or 7-point) scale: 1 if similar, 5 opposite, 2 if almost similar, 4 if almost different, and 3 designates a neutral value or both. Analysis techniques can be used to build other matrices from the grid (e.g., a matrix of *element by element*, or *construct by construct*).

A simple calculation of similar scores by element will reveal those elements with the highest value construct. More involving is the cluster analysis or others. The intent is to establish a criticality value to severe exceptions and/or aberrant decisions. The collection of all exceptions and the collection of all decisions can be in a hierarchical or networked structure. This is elaborated on in Chap. 5.

Oversight Organization Unit for Control

As a check and balance we need to watch the decision's made by executives and management, since they might exercise abuse, greed, power, etc.—as we have seen in Chap. 2. To this end, the *Oversight organization unit* is charged by the Board of Directors.

This unit operates independently alongside the institution organization structure. The Oversight organization unit is the add-on analogous to the immune system with functionality enhancement. It can be simply committee-based and/or hierarchical. This add-on structure is similar to Congress or Parliament in government, the Academic Senate in a university or the Faculty Council in schools or colleges.

Its responsibility is more than being a watchdog. Observe that this unit can be seen as an upgraded and enhanced unit from the commonly known crisis management unit which is created when a crisis occurs in an institution. This topic will be elaborated on in Chap. 6.

SUMMARY AND DISCUSSION

In summary, we follow a systemic approach towards a basis of a theory of prevention, which at this time is more empirical than theoretical. The approach is based on the biological spectrum. We claim that the scope of the biological spectrum-based framework houses any business problem and the formulation of an σ-algebra of exceptions-decision complex as a measure embeds a solution. The argument is that humans, as part of nature and the biological spectrum, create products (business, industry, or government) using their knowledge, experience, and decisions (Fig. 3.1), which all belong to the biological spectrum, physically and/or mentally. One way to discredit this claim is to find a single example of past problems or solutions which does not belong to the framework.

We defined an analogy-based model (Fig. 3.2) which fosters the basis of a theory of fiasco prevention, focusing on exceptions, decisions, and control. We proposed two aspects of analogy. The first goes across all component levels of the biological spectrum and therefore addresses the breadth of the problems and potential solutions. The second covers the three layers of pairwise analogues: guiding principles (top layer or strategic); organization (middle layer, structural, functional, and behavioral); and supporting entities (bottom or operational layer). This addresses the depth of the problems and potential solutions. We use facts drawn from past bankruptcy cases to justify of our systemic approach.

The analogy between cell-human, employee-institution, institution-market, and market-economy (as shown in Fig. 3.3) allows us to identify some anatomical entities (structural) of cells analogous to humans. Examples are entities such as macromolecules, blood, cellular exchange, and interstitial fluid, governed by human DNA/genes as analogous to entities of human-employees, such as projects, funding, transactions, and data, all governed by the institution's regulations and rules.

Three interrelated application domains are suggested and sketched out in this analogy towards a tetrahedron model for fiasco prevention (as shown in Fig. 3.5). The first involves the formulation of exceptions management as a revised, integrated, and enhanced implementation of a modified ERM or MBE system, targeting the business supporting entities. The second involves a psychology-based decision model for understanding decisions adapted from George Kelly. The third involves the management control of the intertwined exception-decision by an Oversight organization unit.

We acknowledge that any bankruptcy is extremely complex, before and after. Even more than 13 years after Enron, after more than ten congressional committees on the case, a huge amount of literature, reports by professionals in all related disciplines, with many lessons learned, a few dozen Enron-related bills introduced, and additional reforms, we are still learning from the various cases. The whole story on executive decisions cannot be fully understood due to responsible parties claiming the Fifth amendment. We need to find the best way to work with what we can find out and place it in this biological framework towards a mathematical formulation as an σ-algebra measure space for anticipating and preventing fiascos.

CHAPTER 4

Preventing Corporate Fiascos: *Corporate Information Exceptions*

In Chap. 3, under the heading of conceptual modeling of exception management, we proposed at least three issues should be considered. We need to detect exceptions (signs and symptoms), a *monitoring issue*. We need to expose exceptions, as they occur, to all responsible parties, *a transparency issue*. And of course there is the *validation* issue, since we need to know whether the exceptions are valid. After validating the exception, we associate it with a *criticality* value, which is subjectively assigned by the employee in charge.

We can analyze various corporate fiascos and use as cases to detail requirements definition for the exception management system. We can then proceed with the design of an application to meet these requirements. As we introduced in Chaps. 2 and 3, we will use, in addition, the analogy and analogical reasoning of cancer between cell-human and human-institution to design the appropriate features for this application.

One objection to the idea of using a cancer analogy for insights into the detection of fiasco symptoms may be due to the following reasons. First, cancer symptoms are extremely difficult to detect. Second, by the time the signs or symptoms surface, or are discovered, it is commonly too late since cancer has proliferated to other organs in the human body via the blood circulation and/or lymphatic system. Third, cancer in humans is a deadly disease, so why would we want to use it as an analogy?

Our reply to this objection is based on a very simple fact. Early signs and symptoms are there, one way or another, if a tumor starts to grow, benign or malignant. They are only detected by the autonomic nervous

© The Editor(s) (if applicable) and The Author(s) 2016   57
T.N. Nguyen, *Preventing Corporate Fiascos*,
DOI 10.1057/978-1-137-49250-0_4

system below the consciencious level. Biologically and medically, we have enough understanding of how the signs and symptoms of a malignant tumor show themselves in the interstitial fluid between cells and in the blood plasma below the conscious level. Currently, we just do not have the sophisticated technology (Albert et al., 1998; King, 1996) to find early enough these symptoms at the conscious level

The symptoms are difficult to detect because the growth and invasion of a tumor develop under the control of the autonomic nervous system, i.e., at a level lower than the human consciousness. Therefore the human in question is unaware of them.

We need to go below the level of consciousness to see what is happening in the interstitial fluid (IF) between cells. This can include, but is not limited to: (1) how the IF pressure and IF flow are involved in the stiffening of nearby tissues; (2) how the malignant tumor generates its own blood supply; (3) how the cancerous cells escape detection as foreign cells in lymph nodes; (4) how the T-cells and B-cells of the white blood cells fail to kill cancerous cells at these nodes; and (5) how these cancerous cells reenter the bloodstream to proliferate in other organs and organ systems in the human body.

We want to see whether we can equate the above at the cell-human level to that in the human-institution level. That is, we mimic the functionality in the cell-human component to introduce an improved functionality in the human-institution component. But first let us analyze the basic requirements using Enron as use case.

Enron Bankruptcy as Use Case

Enron Collapse in Brief

On February 2, 2001, Enron filed for bankruptcy protection (Fox, 2003). Its stock fell from $90 per share to some 20 cents. Three months earlier, Enron posted a huge loss in its third quarter financial statement. This triggered a review of financial statements for the three previous years, causing the stock price to drop to the tens of cents. The following briefly outlines what happened and when (Heally et al., 2003). A statement in *italics* at the end of each numbered item reveals the initial requirement for detection.

1. Enron, the former Houston Natural Gas, was already in debt after a merger with InterNorth, a gas pipeline company of Nebraska. CEO Ken Lay hired Skilling in 1990 to serve as a McKinsey & Co. con-

sultant to Enron to develop a more competitive business strategy. Skilling promoted his *mark-to-market* (MTM) strategy to Ken Lay with the concept of a *gas bank*.

Note that MTM was not new to accounting or financial markets. It has been a means to adjust fair market value, recommended by US GAAP (Generally Accepted Accounting Practice) and approved by US SEC (Securities Exchange Commission). However, the use of MTM as a trading model, as suggested by Skilling, was new in the sense that it was an energy derivative where Enron would act as intermediary "bank".

As such, Enron assumed the risks in purchasing gas from suppliers and selling to consumers at contractual fixed prices and service fees. MTM was used to achieve two successful projects: Cactus 3, with GE Credit Corp and other banks as partners; and JEDI, with CalPERS partnership (California Public Employees' Retirement System). *Other projects which followed JEDI were troublesome but they went undetected.*

2. There were other strategies, besides MTM, such as the shift from asset-heavy to asset-light—nicknamed Death Star, Load Shift, Get Shorty, Fat Boy and Ricochet used in California's energy market—and a diversification strategy to businesses other than gas and electricity. The strategies were realized in many subsequent projects: the electricity project Dabhol in India; a gas project in Bolivia; a water project with the creation of Azurix Corporation; a broadband project with Blockbuster Video; and more. These projects were funded by many special purpose entities (SPEs) created to minimize investment risk as off-balance sheet in partnerships with Chewco LLC, JEDI II, LJM Cayman, LJM 2 and LJM 3, Braveheart, and Raptors during 1997–2000. *These strategies were questionable.*

3. To further support the MTM strategy and others Skilling set out to hire the best and brightest traders. He devised a new policy on performance review, called *360-dregree review*, and a new mantra, respect, integrity, communication and excellence (RICE). The policy appeared great in theory but its implementation departed from common human resource practices (e.g., no job description). The main goal was to motivate traders to initiate any deals that had the potential to generate a good return. The hiring was not limited to traders but included auditors and counsels. Even with an attrition rate of 15 % or higher, Skilling's team was able to grow the set of special revenue-driven employees to act Skilling-like. One of whom

was Andrew Fastow, hired by Skilling in 1990. *The policy was an alteration from human resource standards.*

4. Skilling successfully convinced the internal counsel, the Andersen accounting audit, and the SEC to approve the MTM change to accounting practice (Ackman, 2002; Pollock, 2002; Reinstein & Weirich, 2002). With increasing success, after becoming Enron COO in 1996, Skilling applied, with Ken Lay's approval, the mark-to-market trading model and accounting practice, beyond the gas business. He expanded to other businesses such as electricity, coal, paper, steel, water, weather, and so on (Chatterjee et al., 2002). He appointed Fastow as CFO to oversee Enron financing in 1998. The Enron board of directors also approved Fastow's dual role as Enron CFO and SPE manager in 1999. Taken together, the MTM trading model, the group of top traders, the expansion to all other businesses, and the bull market during the 1990s which facilitated investment opportunities, resulted in an incredible Enron success. *There were violations of Enron's business code of conduct.*

5. Since 1997, however, Enron profits were squeezed due to new entrants and other smaller competitors, including Dynegy, Duke Energy, El Paso, and Williams. Enron began to lose the competitive advantage. To be financially able and to maintain high credit ratings, Enron devised the use of SPEs to access capital and hedge risks as they entered into new mergers and acquisitions. *The company became more of a hedge fund than a trading company* as observed by C. W. Thomas (Thomas, 2002).

6. SPEs were shell partnerships sponsored by Enron, supposedly funded by independent financing (Holtzman et al.. 2003; Jickling, 2003). Two conditions should have been satisfied to keep the SPEs separate from Enron: at least 3 % equity; and 50 % or more control of financial interest given to independent investors. The SPEs were used to purchase forward contracts with producers and sell under long-term contracts to consumers. However, the conditions were not always clear. In the Chewco partnership, for example, 3 % of JEDI II SPE was owned and actually controlled by Enron executives. The situation was much more complex in Raptors. *These were signs of fraud.*

As reported in the literature, Andrew Fastow, the mastermind behind all the SPEs, used some 3000 of them as a way to transfer losses off Enron's books and to book them as revenues. He used them to maintain and/or increase credit ratings by reducing debt-to-total-assets ratio in projects, namely Chewco LLP, LJM Clayman, and LJM 2. As later discovered and reported in court documents a total of 22 % of Enron expenditures from 1998–2000 were write-offs, starting from October 2001, $287M for Azurix, $180M for broadband with Blockbuster, and $544 for others. In addition, Portland General Corp was sold for $1.9B at a $1.1B loss.

How It Finally Happened?

The first official exception to capture everybody's attention was in the press release of Q3 2001 on October 16, 2001. There was no balance sheet disclosure. Also, it revealed a $1.2B charge against equity. About one week later, SEC inquired about SPEs. Then, facing pressure from Wall Street, in November 2001 Enron admitted it had hidden losses in the SPEs and posted re-statements with consolidations for the previous years (1997–2000).

The re-statement of financial reports in the third quarter of 2001 and the financial consolidations resulted in Enron stocks dropping from $90+ per share to 26 cents per share within two months. It brought the company to collapse in November 2001.

What Has Been Done?

Since the collapse of Enron in 2001, there have been *lessons learned* reported by numerous researchers and professionals, and *solutions and recommendations* for the purpose of preventing another Enron (Bierman, 2008; Deakin & Konzelmann, 2003; Dharan, 2002; Higgs, 2003; Nakayama, 2002). Yet a series of other bankruptcies followed, as we selectively detailed in Chap. 2: WorldCom in 2002, Adelphia in 2003, Parmelat in 2003, and one of the biggest bankruptcies, Lehman Brothers in 2008.

Without repeating the details of the above collapses, one obvious observation was that they were mostly caused by fraud. One could ask:

- Were the solutions and recommendations too late for those institutions after Enron since the fiascos were rooted too deeply?
- Although the solutions and recommendations were all good, were they collectively not enough to prevent the collapses from happening?
- Did the institutions (i.e., management) just ignore them?

Finding the answers to the above questions is part of this work. The Enron case, with the initial requirements identified in *italics* in the numbered list in the previous pages, together with the Barings and Lehman Brothers cases in Chap. 2, when combined for a requirement analysis, are a good starting point for a design problem.

It helps to identify at least three features: (1) data collection; (2) data analysis for identifying critical signs and symptoms as exceptions; and (3) exception reporting to all responsible parties for decisions and actions. The difficult question is *how*, especially for the data collection and analysis features.

In searching for an answer to the *how*, in part, all the lessons learned and analyzed in various domains—accounting, legal, financial, governance, etc.—and made available in the literature were used as criteria for detection. In addition, we need to mimic the autonomic system's functionality of symptom detection in the cell-human component for application in the human-institution component. To that end, we need more insights from the cancer analogy for an improved exception management.

First, we look at the details in the human's *internal environment*, which, in essence, consists of the interstitial fluid and the blood plasma (Gross, 1996). This is considered analogous to the *data environment* found in any institution, which all employees deal with on a daily basis. Next, we identify factors which involve (1) cancer growth and development in tissue invasion, and (2) cancer proliferation to other organs via blood circulation and the lymphatic system (King, 1996). We would like to equate the signs and symptoms of cancer disease occurring at the interstitial fluid level of the cell-human component to those of a fiasco in the corresponding level of the human-institution level.

Exploiting the *Cell-Human* and *Human-Institution* Analogy

Detailed Analogy Between Milieu Interieur *in Human and Data Environment in Institution*

Recall that we have sketched (Fig. 3.3, Chap. 3) within the data environment, the operational, managerial, and strategic activities as projects, which are biologically analogous to macromolecules produced by cells (described in Fig. 4.1). These projects are budgeted and funding is allocated. The funding is considered analogous to the blood supply to cells.

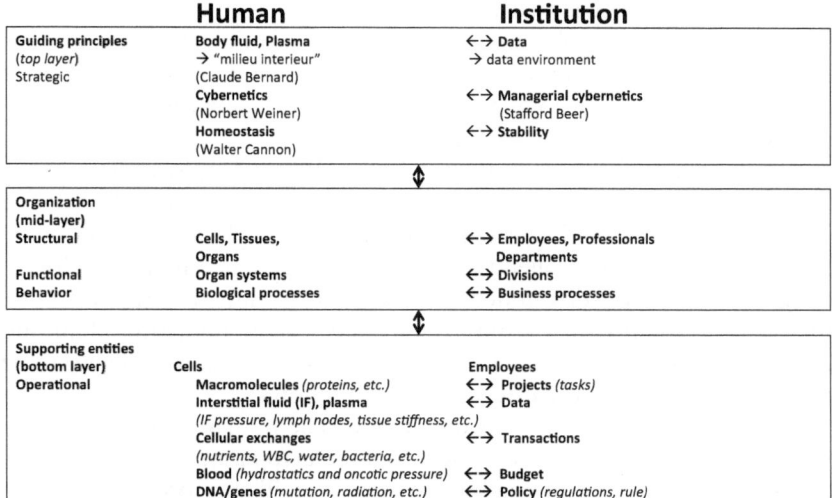

Fig. 4.1 Analogy between cell-human and human-institution components

The institution needs funding to pay and to support employees to do their work, much like the human needs oxygenated blood for the cells. The blood is supplied to cells for their survival and growth. The data are supplied to employees of the institution via numerous information systems and databases/data warehouses for them to carry out their daily tasks.

Transactions between employees in the same or different projects occur. This is similar to exchanges between cells—the amounts or volume of exchange or release of chemicals, water, proteins, white blood cells, etc.—in different tissues—across cell membranes, between capillaries, and interstitial fluid.

And finally, all of the above are governed by the genes in DNA, much like all human entities are governed by institution policies, such as regulations, rules, systems operating procedures, etc. In fact, policies (similar to DNA/gene expressions) are statements that govern the above four corporate supporting entities. The associated policies can be specific, detailed, and measurable at the lower level of the organization. There can be lower-level processes/procedures defining anything and everything, including evaluating a loan application, starting up a server, making a reservation for meeting room. A difference between human DNA/gene expressions, which are mostly readily available in their original form unless mutated, and institution policy is that the latter can grow from simple procedures and become more complex and mature as the institution grows. The policy could then become sketchy, unstructured, and robust at higher corporate levels. Details of the analogy of supporting entities between the two environments are further given in Fig. 4.1.

A cell is the basic unit of a human body. Human-employee is the basic unit of an institution. As noted, the cells in a human body live in the internal environment—i.e., the interstitial fluid and plasma, which Claude Bernard (Gross, 1996) called the *milieu interieur*. The human employees of an institution are surrounded by its data environment in the same way that cells are bathed in the interstitial fluid . Since interstitial fluid and plasma are practically the same in terms of constituents, we only examine the interstitial fluid in the discussion which follows.

The cell is surrounded by its membrane. The human is surrounded by skin. Each cell consists of the nucleus and the cytoplasm. The nucleus of a cell is *similar* to the brain of a human employee. The cytoplasm, in which organelles exist, is *similar* to the organs of the human employee body.

Macromolecules (i.e., proteins, carbohydrates, lipids and nucleic acids, etc.) are produced in the cells, much like projects are carried out by employees in the institution. Each protein is a chain of polypeptides consisting of amino acids, much as a project is a series of tasks to be performed. There are numerous types of proteins in a human, just as there are multiple projects in an institution. An example is the protein receptors of different kinds in humans, which are comparable to the data collections of different kinds in an institution.

The human body works miraculously to maintain homeostasis, from the cell as the unit of life, to all tissues, organs, and organ systems. Note that the homeostasis scheme comes into play when excessive or insufficient nutrients or debris are detected by cells playing the role of chemical sensors, e.g., alpha cells and beta cells. Any change in the human body is detected by numerous internal mechanisms. We are particularly interested in the autonomic part of the brain which operates under the consciousness level to maintain homeostasis. For example, baroreceptors can detect a change of pressure in the bloodstream passing an artery or arteriole to help the latter dilate properly. The following explains further.

Stability in Institution and Homeostasis in Human

Consider blood sugar as an example of homeostasis at work (Cannon, 1963). Either low or high concentrations of glucose in the blood are dangerous to the body. Control is performed by the alpha and beta cells in the pancreas to maintain a narrow range of blood sugar. Low concentrations will be helped by the amount of glucose released by the liver into the blood. High concentrations will be neutralized by the release of insulin to remove excess glucose to maintain homeostasis.

Another example, adapted from Walter Cannon's *The Wisdom of the Body* is presented here. In normal operations, the cardiovascular portion of the circulation system provides oxygenated blood and nutrients that exit the left ventricle to the aorta. The blood travels to the arteries and arterioles to arrive at all cells of the body bathing in Claude Bernard's *milieu interieur* via the mesh of capillaries in every tissue. It is from this *milieu interieur* where debris, dead cells, waste, including cancer cells, are picked up by the tiny lymph channels. These are exchanges via changes in concentration and pressure, much like what happens in the capillaries. The latter deposit them to lymph nodes, which are then dumped in the veins to return to filtering organs (liver, kidneys) and the heart. The cardiovascular system and the lymphatic system work together within a global and local signaling system.

The signaling system involves sensors (Wiener, 1948). Examples of sensors are receptors for pressure differences, receptors for concentration differences. The system also involves the brain, spinal cord, gland systems, and so on, with hormones and impulses. Via these systems, information is integrated and transmitted among all parts of the body to address external and internal changes.

This is how, in general, the human body is able to maintain a body temperature of 98.6F despite changes of temperature in the environment. This is how tumor suppressor genes in cells can destroy cancerous cells at their first occurrence. This is how the defense system of T cells and B cells in the blood react to abnormal cells in the body. These cells are lymphocytes which produce antibodies. And this is how all organs and organ systems work together to keep the internal environment constant.

When events are out of bounds, as detected by the associated sensors such as alpha or beta cells in the spleen, corrections based on regulations (e.g., gene regulations or autonomic nervous system of the brain) occur to maintain homeostasis. The inability to make corrections will result in exceptions (or symptoms) expressing abnormal, deviated behavior, such as those caused by cancerous cells and malignant tumors.

The concept of business stability in institutions, analogous to the homeostasis concept in the body, has received little attention from institution management. The reason is that the major objective in these institutions is financial gain, not stability. Formulating a business stability-driven criterion allows the identification of non-stable events or exceptions caused by activities within the institution. These exceptions, as signs or symptoms of problematic events, could identify the aberrant decisions exercised by a group of employees at any level of management, as in the case of Enron.

Enron as a Cancerous Institution

We have mentioned that when an institution is considered as analogous to a human then the institution's employees could be considered as biological cells. Leading to institution tumors being viewed analogously as groups of employees in organization units which grow uncontrollably, and may become *malignant*. They can expand to other organizational units and become seriously harmful to the health of the institution, leading to bankruptcy.

We described Jeffrey Skilling, Andrew Fastow, and other Enron executives as "special" or "abnormal cells" in the institution body. Skilling used an accounting practice which he convinced the SEC to approve after he officially joined Enron. In a sense, Enron's MTM accounting practice was a mutation to common accounting practices, similar to a mutated DNA/gene of a cell, therefore affecting its growth (cell division). Enron's 360-degree review is analogous to a *mutation to human resource hiring/firing policy/process*. SPEs and accounting schemes can be viewed as *changes to or deviations (mutations)* from commonly practiced SPEs and GAAP (General Accepted Accounting Principles).

When Arthur Andersen, the audit arm to Enron, was influenced in mishandling accounting audits (CNNMoney, 2002; Raghavan, 2002), the institution "tumor" proliferated to other institution "organs and organ systems" (finance division, accounting, legal counsel, etc.). Cancerous symptoms started to surface: Lay stepped down; Skilling replaced Lay as CEO; and then Skilling's resignation after only 6 months, forced Lay to step back in. These were followed by the cancellation of deals with Blockbusters and later with Dynegy (Powers, 2002). Following SEC inquiries and the change of management, the revision of financial statements from 1998–2000 was initiated (EnronAnnualReport1998, 1998; EnronAnnualReport1999; 1999; EnronAnnualReport2000, 2000). Discoveries of exceptions in SPEs required Enron statements to include Chewco's consolidations, as well as those of other projects. Enron stock prices slipped. The consolidations showed Enron debts and liabilities previously off balance sheets. These caused additional stock price slips and cash shortfalls. Within just 60 days, the company's stock went down to 26 cents per share. Enron filed for bankruptcy protection.

IMPROVED MBE MODELING CHALLENGE

There is an MBE (Rickett & Nelson, 1987) in some shape or form in an institution. An improved MBE can be built on top of existing applications, whether they are a transaction processing system, a management information system for tactical purposes, or an enterprise information system or decision support system in support of top management. As such, our MBE application records could be meta-records (records of records) which will be created as we need them. Thus, the improved MBE does not interfere with existing applications in the institution other than interfacing them. We elaborate a number of concepts applicable to the building of the improved MBE (Nguyen, 2013, 2014) as follows.

A Model from the Perspective of Information Exceptions

Our thinking is specifically on *how the symptoms potentially leading to collapses can be exposed in time or early enough so that prevention can be exercised.*

We have detailed the bankruptcy cases from a different perspective, that of a cancer. The analogy to cancer requires us to re-examine the human body in terms of its anatomy (structure), physiology (functionality), and behavior supported by lower level processes and entities, within the guiding principles of its biological systems.

MBE Revisited for Prevention: Completeness in Information Environment

We established previously that the *data environment* is the counterpart of the *milieu interieur* concept. As we have said, information would be derived from data obtained from the business entities performed by the employees (funding, projects, transactions, accounts, and others), much as the blood and macromolecules exchange across cell membranes and chemical products are produced by the cells.

We make the assumption that *all* data obtained would be online and accountable. If the current MBE system does not have them available, the improved MBE will create the records. The data can be about a hiring/ firing process, a particular job application, a schedule for work, an office supply purchase, an account receivable, a travel expense reimbursement, a plan for manufacturing, strategic minutes of meetings, a document from legal counsel, a product design, financial reports, a negotiation contract, and the like. The data environment is therefore the complete collection of data.

The complexity is that we have data in all possible formats. For an existing institution, they are all there in some format and storage media. Most of them are available for the operational, tactical, or strategic planning of the institution. Most of them are already connected logically by topic, content, or in existing applications. Special documents or artifacts, if needed, can be newly created, collected, reviewed, worked on, etc. This raises the possible need for middleware to incorporate data to the improved MBE.

If anything causes any concerns to the responsible employees, the employee can tag a value to a field in the online *metarecord* of the MBE. This is an *exception* field with a *severity* (or criticality) level attached: normal, *warning* or *severe*. A notification is sent to the appropriate channel if the level is warning or severe. The severity when marked on the record is date/time stamped, and can contain or link to the information on *responsibility, accountability* and *authority* of the employee responsible. The next time the record is called for (by itself or part of a chain of linked records), the exception flag will appear along with other data or information. The chain of records is obtained, either via integrated (hierarchical) or aggregated (across hierarchies) schemes.

MBE Revisited for Prevention: Automated Policy Hierarchy and Processes

As the backbone of the MBE, an institutional policy system has to be in place to account for all policies and regulations. Policy in an institution acts like genes in our body. Most employee tasks are understood for day to day operations. The related policy (regulations, rules, etc.) and processes have been guided by well-thought out and well-defined policies in the institution for years.

A new policy of a strategic nature (or corporate strategy) is more complex. It is hard to detect a flawed policy, especially when the policy or strategy shows some initial success. In fact, they are often unstructured, unclear and the effects are not known until the execution of the policy (or strategy). An example is Jeff Skilling's mark to market. It was a modified strategy to promote a gas bank business which did generate two successful projects (Cactus 3 and JEDI), but it started to go bad.

Deviations from current policy or strategy, if detected, would be flagged. As in the case of Enron's 360-degree review, which departed from common human resources practices. This is the case for the MTM strategy adapted for gas bank and expanded to other business. This is the case for MTM accounting, which was "mutated" from US GAAP. This is the case for SPEs, where the 3 % condition in SEC regulations was violated. This is the case where a dual role for Andrew Fastow was allowed by the Board of Directors.

MBE Revisited for Prevention: Complementarity Property

Both human body and institution observe a hierarchical organization. However, the human body hierarchy is a little different from the institution's hierarchy. In the human body organization, there are multiple paths from a lower level to higher levels.

In fact, the atoms, molecules, and macromolecules are not necessarily connected only to an immediate higher level. They can connect to other higher levels directly. This type of hierarchy enjoys the *complementarity* concept according to Rick Looijen (Looijen, 2009), which says that *the whole is complemented by all of its parts*. As pointed out by Looijen, this path organization may increase the complexity but it promotes the excellent functioning of the ten organ systems of the body in support of the biological processes. Institution organizations, either strictly hierarchical (such as common businesses) or committee-type (such as the US Congress or a university) do not entertain this type of hierarchy with complementarity.

We would allow the complementarity feature in our implementation model. Entities such as funding, projects, transactions, and accounts are linked to express the *complementarity* property. This is achieved by allowing lower level institution entities to link to all higher-level entities, and sideway linkages between entities on the same level of organization (or level of abstraction). This linkage is implemented as links with *roll-up* and *drill-down* capabilities, observing the inheritance property (integration), and/or *sideway* links, observing the aggregation of different entities.

Higher-level entities (for examples projects of projects, transactions of transactions of projects), as parent or sibling entities, would include all exceptions that occur from lower-levels. Integration is done for each hierarchy of relevant nodes. Aggregation is done for nodes that are parts of different hierarchies. Thus, the links are nodes of a network of connections from which their severity is aggregated and integrated. We therefore can drill-down or roll-up within the hierarchy, and offer links among different hierarchies.

MBE Revisited for Prevention: Overall Corporate Governance

It is recognized that both the business market (Moore,1993, 1996, 1998; Iansity & Levien; 2003) and its corporate governance (Ravn-Jonsen, 2009) are extremely complex. In the cases of Barings Bank, with Nicholas Leeson, and Lehman Brothers, involving a group of top executives, corporate governance was identified as one of the key reasons contributing to the institution's failure. An improved MBE must observe the overall corporate governance. This includes that recommendations from the Higgs and Powers reports, as well as those recommended in US congressional reports, are captured as policy. It also means that exceptions are allowed to be overridden by the management-executive team with stated reasons. However, the responsibility of monitoring and reporting must be assigned to an organization unit which is outside of the management-executive team. We call it an Oversight organization chartered by the institution's board of directors.

EXAMPLE IMPLEMENTATION

Enron Case in the Context of Fiasco Prevention

The first and main strategy introduced to Enron by Skilling was the mark-to-market strategy/accounting practice approved by the Board of Directors. This strategy and accounting practice, however, raised concerns amongst some Enron employees. First was David Woytek. In 1992 Wortek questioned the MTM (Eichenwald, 2005) when it was first introduced internally by Skilling's team. His concerns and objections were overridden by Ken Lay and ignored.

When the 1992 annual report was published, the mark-to-market scheme in the footnotes caught the attention of Toni Mack of Forbes (Mack, 1993). She voiced her concerns for the 1992 financial statement footnotes in her "Hidden risks" article in 1993. Again, Ken Lay became angry, replied to Mack personally, and overrode the exception by ignoring it.

In 1999 Carl Bass also objected the MTM accounting scheme but his objections went nowhere. A warning from Sherron Watkins did not go far either.

If the concerns and objections above were properly registered, as dictated in our model, all responsible employees would be aware of what was going on from Day 1. Transparency of the entities would be preserved among the responsible.

More specifically, David Woytek's and Carl Bass's *objections* to the MTM *policy* and the *doubts* raised by Toni Mack in her Forbes article are of the *others* entity type. These activities and events would be marked with the *warning severity*. The objections and doubts (*others*) overridden by Ken Lay would be recorded as "override".

Other examples would include: (1) the *approval of the Board on the dual role of Fastow* in Enron and partnership as policy; (2) the *$30M paid as management fees* to Fastow for LJM and Raptors to appear in *transaction and account*; and (3) the *off-balance sheet amounts* in financial statements in Enron and in SPE accounts and across them, although legally the assets and losses are off-balance sheets (Li, 2010). They should be marked with a severe level of criticality. These indicate potential exceptions subject to correlation in analyses and other means on aggregated records and in the vertical and horizontal integration of information. If aggregated records across all entities (i.e., funding, projects, transactions, accounts, and others) can be shown, it will be harder to hide any of the above. The 3 % *investment of Michael Kopper* (Enron employee) in the Chewco project should have been exposed.

A similar scenario happened in the use of SPEs when Duncan was approached by Fastow to set up an SPE for LJM to be managed by Fastow himself (Fowler, 2002). Although advised against by Benjamin Neuhausen of Andersen Professional Standard Group due to a conflict of interest, Duncan said he would go along with the setup if Skilling and Enron's Board approved. The bad news was that it was approved by both Skilling and the Board. Had these symptoms been recorded officially, via records in an MBE, the chances of them being ignored or approved by the Board would have been much lower. Had an integrated record been shown, the attrition rate (15 % in Skilling's division) should have raised concerns about the effectiveness of the 360-degree-review policy. The huge number of SPEs, in the thousands, should have raised concerns.

Note that detecting exceptions which could potentially lead to fiascos at all levels is not an overnight task, especially when the seeds for later failure are initiated at the top level of the institution, as in Enron case. Note also that at higher-levels of management, most entities are unstructured, especially policy or strategy. The effectiveness of a new policy or strategy—e.g., MTM, virtual integration, diversification, etc.—on a new market, as in the case of Enron—e.g., energy "bank", energy trading, commodities trading, etc.—cannot be seen until it is at least partially implemented. It is also more difficult to discover any wrongdoings when the said policy or strategy, such as MTM, initially works successfully.

As previously alluded to, we are not only interested in *why* Enron collapsed (e.g., conspiracy, greed, corruption, abuse, ethics, and others) but also in *how* Enron exceptions occurred, to trace the symptoms in documents and artifacts based on what we now know. We may want to know, however, *why* the exceptions were not caught until October 2001, despite some early and sporadic warnings and explicit objections—e.g., fuzzy accounting, clear conflict of interests, etc.—between 1992 and the fall of 2001. This type of understanding is necessary.

Another idea is about a mechanism to make the symptoms transparent in the MBE. The symptoms are definitely buried somewhere in the business entities of funding, projects, transactions, accounts, and financial reports. They are either not discovered or ignored, or forced to be ignored by the responsible people—e.g., the board, auditors, and counsels—despite all rules and regulations. .

Thus, in brief, for detection of exceptions as signs and symptoms an improved MBE has two basic features:

1. The ability to capture data on five entities as metarecords: *policy, data, projects, transactions, funding* and everything else (*others*) with a marked criticality level. Data must be treated as a whole, integrated and aggregated across all functional areas in a transparent and "autonomic-like" system. Data with "warning" or "severe" severity level must be disseminated to responsible people for timely decision making: employees and management, external auditors, counsels, regulators and investors/partners.
2. The ability to do an analysis for a deeper understanding. This also means that the results of the analyses will be disseminated to the proper authority, and the ability to allow analysts to voice their findings of discrepancies or exceptions. Again, it must be transparent to all responsible people, and therefore any overriding actions by an authority are accounted for.

Feature (1) results from the principle of *data environment*, while feature (2) supports the principle of *managerial cybernetics* for *business stability*.

In the example illustrated below (captured in Fig. 4.2), we showcase Enron exceptions in an improved MBE, revealing symptoms leading to fiascos as proof of concept.

Figure 4.2 Explained

The first row (P-003) shows SPE as a policy (or strategy) devised by Andrew Fastow to enter into a partnership with investors. The second row records as a project Jeff Skilling's decision to invest in Rhythms NetConnections, a high risk business (therefore originally marked as *warning*). He approved the investment nevertheless and turned it to *normal*. When asked by Skilling, Vince Kaminski performed a financial analysis and concluded it was severely risky (third row—marked as *severe*).

The severity shown by the collection of *warning*s and *severe*s would have triggered the attention of the responsible parties even though some indicators were overridden.

As part of the scenario, LJM Cayman was drawn to hedge Rhythms NetConnections (4th row, *warning*). Fastow used Enron stock and his own investment to go into partnership with the Cayman Islands and overrode the warning (5th row, *warning*). Both were high risk decisions, however overridden by Fastow to turn them to *normal*. The project and its transactions were carried out anyway to hide the losses in this investment in the 1999 financial statement (6th row, turned *severe* to *normal*).

Other projects, accounts, transactions and financial statements followed (not shown) until the financial statement of 2001 3rd quarter (8th row, *severe*). When SEC inquired (9th row, *severe*) about the obscure 2001 financial statement of the third quarter, other financial statements of

ID	Related entity	Entity type	Date	Severity	Link to	Action by	Override Date	Reason	Severity
P-003	SPE creation	Policy	1991	normal	PR-99	Fastow	1999	Partnership	normal
E-191	Rhythms Net	Project	1999	warning	PR-99	Skilling	1999	Investment	normal
E-192	Financial analysis	Task	1999	severe	T-099	Kaminski	1999	Risk	severe
E-193	LJM Cayman	Account	1999	warning	A-099	Fastow	1999	Hedge	normal
E-194	Cayman Islands	Transaction	1999	warning	TR-99	Fastow	1999	Enron stock	normal
E-195	FS-1999	Others	1999	severe	O-199	Skilling	1999	Statement	normal
...	...	...	...		...	...	...	...	...
E-211	FS-2001 Q3	Others	2001	severe	O-211	Skilling	2001	Consolidation	severe
E-212	SEC inquiry	Others	2001	severe	O-212	Fastow	2001	Consolidation	severe

Severity is gray scale coded: white means *normal*, light gray *warning* and darker *severe*

Fig. 4.2 Enron Rhythms NetConnections project

1997–2001 were restated with losses (not shown). This eventually led to the collapse.

The list of exceptions in Fig. 4.2 was compiled from available sources (policy, project, funding, transaction, accounts, and others) for the improved MBE with detailed information on who did what, where, when, and how in supporting documents as captured partially in Figs. 4.3 and 4.4.

Figure 4.3 Explained

The *SPE creation* (first row in the exception report in Fig. 4.2) was among three strategies listed under the policy entity report (top part of Fig. 4.3). The *financial analysis* task was assigned to a professional (Kaminski) who raised concerns (*warning*). The task entity report included other tasks performed by an Enron manager (Woytek), an accountant (Bass), and an outsider (Mack) who raised concerns (all *warning*) about the MTM scheme in the 1990s. The project entity report shows an example of a good project (Cactus III—marked *severe*) and two other projects (Chewco and Rhythms NetConnections). This report has warning indicators due to SPEs partnerships (*warning*), however both of them were overridden by Lay and Skilling.

Figure 4.4 Explained

Figure 4.4 shows the corresponding reports for accounts, transactions and others from which the symptoms on exceptions were compiled. Every transaction is aggregated in accounts and every task is aggregated in projects (detail not shown). Every account links to one or more projects as siblings. The marked severity level was roll-up from tasks, transactions and others to projects and accounts. The reason for the severity of any entity can be found by drilling down to its constituents. Communications or artifacts are expressed as "Others" if they are not the result of a task

ID	Entity type	Name	Who	Date	Severity	Roll-up	Drill down	Supporting docs	Sibling	Action by	Override Date	Override Reason	Severity
P-001	Policy	MTM strategy	Skilling	1990	normal	N/A	P-002	N/A		Lay, BOD	1991	Parnership	normal
P-002	Policy	MTM accounting	Skilling	1991	normal	P-001	PR-01	Internal doc	PR-01	SEC	1992		normal
P-003	Policy	SPE creation	Fastow	1991		N/A	A-991	Internal doc	PR-01	Skilling	1999	Hedge	normal
ID	Entity type	Name	Who	Date	Severity	Roll-up	Drill down	Supporting docs	Sibling	Action by	Override Date	Override Reason	Severity
T-001	Task	MTM-objection	Woytek	1991	warning	P-001	PR-01	Internal meeting	P-001	Lay	1992	None	normal
T-002	Task	MTM-warning	Mack	1992	warning	P-001	FS-1992	Forbes article	P-002	Lay	1999	None	normal
T-003	Task	MTM-objection	Bass	1999	warning	P-001	LJM Cayman	Internal memo	Others	Lay	1992	None	normal
T-099	Task	Financial analysis	Kaminski	1999	warning	PR-99	LJM Cayman	Internal report	TR-91	Skilling	1999	Risk	normal
ID	Entity type	Name	Who	Date	Severity	Roll-up	Drill down	Supporting docs	Sibling	Action by	Override Date	Override Reason	Severity
PR-01	Project	Cactus III	Skilling	1993	normal	P-001	T-001	Internal report	A-001	Lay	1993	None	normal
PR-97	Project	Chewco	Fastow	1997	warning	P-001	T-097	Internal report	A-971	Lay	1997	None	normal
PR-99	Project	Rhythms Net	Skilling	1999	warning	P-003	T-099	Internal report	A-991	Skilling	1999	Investment	normal

Fig. 4.3 Business entities

ID	Entity type	Name	Who	Date	Severity	Roll-up	Drill down	Supporting docs	Sibling	Action by	Override Date	Override Reason	Override Severity
A-921	Account	Cactus	Fastow	1992	warning		TR-21	FS-1992	P-001	Skilling	2001	None	normal
A-931	Account	Cactus III funds	Fastow	1993	normal	PR-01	TR-31	FS-1993	P-003	Skilling	2001	None	normal
A-971	Account	Chewco	Fastow	1997	severe	PR-97	TR-71	FS-1997	P-003	Skilling	2001	None	normal
A-972	Account	FS-1997	Skilling	1997	severe	Others	TR-72	FS-1997		Lay	2001	FS	normal
A-991	Account	LJM Cayman	Fastow	1999	severe	PR-99	TR-91	FS-1999	P-003	Skilling	1999	Hedge	normal
TR-21	Transaction	Cactus III	Fastow	1993	warning	PR-01				Skilling	1993	None	normal
TR-71	Transaction	Chewco	Fastow	1997	severe	PR-97				Skilling	1997	None	normal
TR-72	Transaction	FS-1997	Skilling	1997	severe	Others				Lay	1997	FS	normal
TR-91	Transaction	Cayman Islands	Fastow	1999	severe	A-099				Fastow	1999	Enron stock	normal
O-199	Others	FS-1999	Skilling	1999	warning	PR-01				Skilling	1999	statement	normal
O-212	Others	SEC inquiry	Fastow	2001	severe	PR-97		FS-2001		Skilling	2001	Consolidation	severe

Fig. 4.4 Business entities (continued)

or transaction. In the Enron case illustration, the entities marked with *warning* and/or *severe* involved the LJM Cayman and Chewco projects with the SPE accounts.

Extension to Other Business Cases

The next issue is "*Can we prevent another Enron with the MBE system discussed in the previous sections?*"

Recall that our model is *employee-centered,* much as an organism is *cell-centered.* Everything in the institution is considered as a product of employee tasks, much as everything in the human body is the product of a cell's organelles as proteins. The product is described not only in terms of tasks but also as projects, transactions, or accounts, and is not limited to other entities the institution might use.

Secondly, the system is *policy-driven* across all levels of an organization, much like a human body being *gene-driven,* to measure results against "faulty strategy, managerial mishandlings, intended wrongdoings, diverted tasks within projects, out-of-the-ordinary transactions hidden in financial accounts and underreporting statements". These are either supported by existing policies and/or regulations, or questionably violate them, as we mentioned in the first section.

Thirdly, the system is geared towards evaluating and labeling exceptions in terms of criticality level as a complete set of relevant symptoms for a diagnostics of what went wrong. All records have drill-down and roll-up capability, and sideways links. Thus, the MBE system is capable of displaying as complete a set of information on a particular issue as possible, including all analyses substantiating it and all actions/decisions for or against it.

The lessons learned from the Enron use case can be applied to other Enron-like institutions, such as WorldCom, Tyco International, Adelphia, and Lehman Brothers. For example, in WorldCom (Brickley, 2003; Lyke & Jickling, 2002) we now know that the two main figures of Bernie Ebbers, CEO, and Scott Sullivan, CFO, constitute the *institutional malignant tumor*. They were involved in mega-deals with over 60 acquisitions during the 1990s (complex growth, potentially uncontrollable).

The main fraud was in accounting. It involved staff who listed operating expenses as assets, therefore the huge losses turned into enormous profits, some $1.38 billion in net income in the 2001 financial statement. They also misused reserve funds. Meanwhile, Ebbers netted some $140M and Sullivan some $45M from stock sales. The detection of exceptions by the MBE would be easier than in the Enron cancer in terms of transactions, with accounts concerning different projects transparently recorded in an MBE.

The case of the collapse of Lehman Brothers (Azadinamin, 2012) is of a different scale and complexity, however. The effect of its bankruptcy rippled through the market, causing: a large decline in the Dow Jones; involving the subprime market and the market in credit default swaps; the misuse of Repo 105; obscure financial statements; the FED; its complex structure; and many others.

The findings were described in a 9-volume, 2200+ page report by Anton Valukas. Investigations into prevention confirmed that the symptoms of wrongdoings were detectable from cash flow reports, but these were ignored by analysts. Investors and auditors either did not understand the statements or did not see the lies hidden by top management and executives, being blinded by a superficial performance. There were many warning signs. This case is worth an in-depth investigation by the MBE.

There is another specific type of problem, such as the collapse caused by a single person in an organization. Typical ones are Bernard Madoff with his Ponzi scheme (La Roche, 2011), or Nicholas Leeson of Barings Bank (Leeson, 1996). The wrongdoings would have been discovered had the symptoms been recorded and analyzed properly, with complete transparency, to the proper authority. In fact, warning signs were identified by Harry Markopolos, who had been investigating Madoff since 1999. Similarly, in the case of Nick Leeson, the head of Barings Security Operations had been warning top management since 1992 about the dual role of Leeson in trading and settlement. These warnings were ignored or forgotten, since they were not part of an MBE scheme or equivalent.

SUMMARY AND DISCUSSION

In summary, in this initial MBE solution, the *data environment* is biologically-inspired by Claude Bernard's concept of an *internal environment*. It offers the availability and transparency of information to responsible employees, who would be aware of the situation at any time, and could detect irregularities or anomalies and report them in the system as *warning* or *severe* at the appropriate level of criticality.

These exceptions are aggregates of lower level exceptions as in the *control echelon* in Stafford Beer's managerial cybernetics. They are escalated upwards to higher levels to the proper authority for decision making. Furthermore, they report accountability to responsible employees at all levels of the structural or functional organization, including those authorities who override warnings or red flags.

When the number of exceptions is numerous, more attention from management will be given. These exceptions exercise the notion of keeping the performance within defined norms. This is biologically-inspired from the *homeostasis* concept introduced by Walter Cannon and Sherwin Nuland. It allows management at all levels to exercise control in driving the performance back to normal when exceptions clearly indicate out-of-bounds situations, which could lead to more serious or catastrophic events.

The MBE illustrated in this paper is just a scratch on the surface of an application system that could truly prevent institutional diseases. It sets the stage for further horizontal control (via association or relevancy between tasks and transactions, projects and accounts) and vertical control (via roll-up and drill-down) of activities and associated events.

Benefits and Values

At this point, one might say that there is no need for the cancer analogy, since the solution as we have proposed it can be reached without it. Truthfully speaking, had no cancer analogy been drawn, this human-centered business model (as opposed to business-centered, profit-centered, process-centered, etc.) would not make sense. The structural property of complementarity and the information environment would be difficult to conceive. The biological mechanisms would be too far reaching, and other concepts, such as homeostasis or cybernetics, would not give rise to a stability with managerial cybernetics.

We claim that the improved MBE model focuses on business *stability* as one of the strategic goals. It is not commonly the major concern of executives. They want profitability. They want growth. They want market domination. They rarely think of stability. They take risks. They become greedy. They commit fraud. We argue that if stability is one of the goals, not only can exceptions be found and fixed but it would be a solid approach to growth and profitability.

So, why all the collapses that followed Enron's failure, despite new laws, rules, and regulations such as Sarbanes-Oxley? The answer is embedded in the next consideration. Human nature spreads between two ends of the dipole concept: conservative and liberal; low-risk and high-risk; right and wrong; restrained and greedy; etc. This is the situation in the business environment. No matter how clever the remedy or prevention one comes up with, there will be someone or some group who is smart enough to break the rules for whatever reason. To fix this, the issue could involve careful hiring, leadership and ethics, and so on, which go beyond the scope of this chapter. Our approach to this issue is to understand human decisions on exceptions which lead to fiascos, i.e., exploiting the somatic functionality of both the human brain and the mind.

Another important point is the institution's management leadership team. Despite all discoveries of exceptions, executives can choose to ignore, override, or hide them. How can we make sure they take a look at the warnings to make proper decisions? We will force each of them to think on the warnings by *eliciting* their input on each and every one. This is to build a set of personal constructs in the sense of George Kelly. The constructs are then *analyzed* and *evaluated* in a fashion similar to Valerie Stewart's revised Repertory Grid, with added considerations on neuro-economics. The combined measures can be expressed on a failure-success scale which measures the management leadership team's performance. This is the topic of Chap. 5.

There is no guarantee that our framework and MBE model will work absolutely, but it can embrace other solutions for the simple reason that it is a shell with metadata on existing data (*funding, projects, transactions, accounts, policy, others*). It can grow as fast and/or be as wide as our means can afford. Philosophically, it seems like a natural way to solve problems caused by humans, which themselves are part of nature.

To our knowledge, there is no other business framework based on insights from the biological spectrum, from *protoplasm to cells-organisms-ecosystem, to biosphere*. None of the previous investigations on preventing another Enron bankruptcy used an approach similar to a cancer or disease analogy. None of them have looked at the conceptualization and implementation of an analogous autonomic feature of the brain to detect the signs and symptoms of fiascos, for a formulation for prevention. And none of them have used features of the lymphatic system to provide facilities for handling signs and symptoms.

CHAPTER 5

Preventing Corporate Fiascos:
Understanding Corporate Decisions

In Chap. 4, we added two important concepts to the conventional management by exceptions system. First is the early discovery of signs and symptoms of a fiasco by implementing a feature in the corporate MBE analogous to the detection of symptoms in the autonomic nervous system of the human body. At cell-human component level, humans are not aware of such early symptoms of exceptions in their body until they become critical because the human body is not designed to do so. There is nothing we can do to change that. But at the human-institution component, we can mimic this functionality and expose exceptions to the responsible parties in an institution. Second is the implementation of human immunization-like features in the MBE.

Nevertheless, the above constitute only a partial solution to the fiasco prevention problem. The other part lies in the hands of executives who make decisions regarding the signs and symptoms, whether they cause them or not. These decisions, if aberrant, could drive the institution into a fiasco.

Given Albert Camus's statement, "Life is a collection of choices", and that choices are made by living organisms with a consciousness (animals), by using both consciousness and self-consciousness (humans), we can observe the following:

1. *Nobody can make all the right decisions all the time and/or fully predict the effect of the outcomes of their decisions.* In other words, humans make mistakes. FED Chairman Greenspan admitted at a congres-

© The Editor(s) (if applicable) and The Author(s) 2016   81
T.N. Nguyen, *Preventing Corporate Fiascos*,
DOI 10.1057/978-1-137-49250-0_5

sional hearing that he had been right 70 % and wrong 30 % of the time over his 21 years of service (Hearing-Greenspan, 2010). One of those wrong decisions was that he did not interfere with the downturn in the subprime market. President Clinton made the decision to facilitate financial services and derivatives in the Gramm-Leach-Bliley Act of 1999 (CLinton, 1999). President Bush made the decision to provide affordable housing to Americans with a low income and low credit score in the American Dream Downpayment Act, 2003(Bush, 2003). They could not have imagined that the many institutions involved in the subprime loan could manipulate the process and the housing market to the point where it could create the turmoil that caused the economic meltdown of the late 2000s.

2. *Most decisions are emotion-driven.* This is confirmed by Antonio Damasio in his famous book *Descartes's Error* and is discussed in Daniel Kahneman's *Thinking: Fast and Slow.* The physical reason is due to the three in one structure of the human brain, called the triune by Paul MacLean: the reptilian brain (fight or flight); the mammalian brain (emotion); and the neocortex (logical thinking) as performed by the prefrontal neocortex and the precuneus in the superior parietal areas of the brain.

3. *No single decision can lead to fiasco.* It has to be a series of decisions over space and time, as seen in the case of many.

4. *The need to understand decision makers' decisions.* Our idea is that, as outsiders, we cannot stop top executives from making their decisions, whether they are right or wrong. That is their responsibility, authority, and accountability, which go with the job, but we can question and evaluate them. The key issue is to question them early enough, i.e., from the occurrence of the first exceptions.

5. *Good people can make bad decisions.* Some might genuinely believe that their decisions and decision making process are sound but their decisions turn out to be bad. From different perspectives their decisions can be extremely good or bad or anything in between. Considered from a decision analysis and quantitative method viewpoint, there are: (i) qualitative or quantitative deliberations of decision alternatives; and (ii) the outcome of the selected alternative. There is the chance that good decisions or no decisions may yield bad outcomes. In most cases, good people would try to correct them.

6. *Some people have a tendency to commit fraud.* People make decisions based on feeling, intention, belief, ethics, and/or other mental abilities. In some cases, due to self-interest or other reasons like greed, people commit fraud. Some whose decisions cause exceptions might try to cover them up, whether or not they know the consequences of this cover-up. CEO Richard Fuld of Lehman Brothers is such an example. Ironically, he re-emerged in the spotlight in 2015 and confirmed that he had done what had to be done.

From our biological spectrum perspective, decision making occurs at five levels, as shown in Fig. 5.1.

We would like to elaborate a little bit on the rather busy Fig. 5.1 The top box sketches the biological spectrum from protoplasm to biosphere. The bottom box is the same spectrum viewed from the perspective of exceptions or anomalies.

In the middle are five levels of hierarchical decision making: cell; human; subject matter expert (SME, collective thinking); corporate (collaborative thinking); and systemic decision making (systems thinking). Each level is a complex network structure (not shown). The human brain has a network of neurons; the institution is a collective and collaborative networked busi-

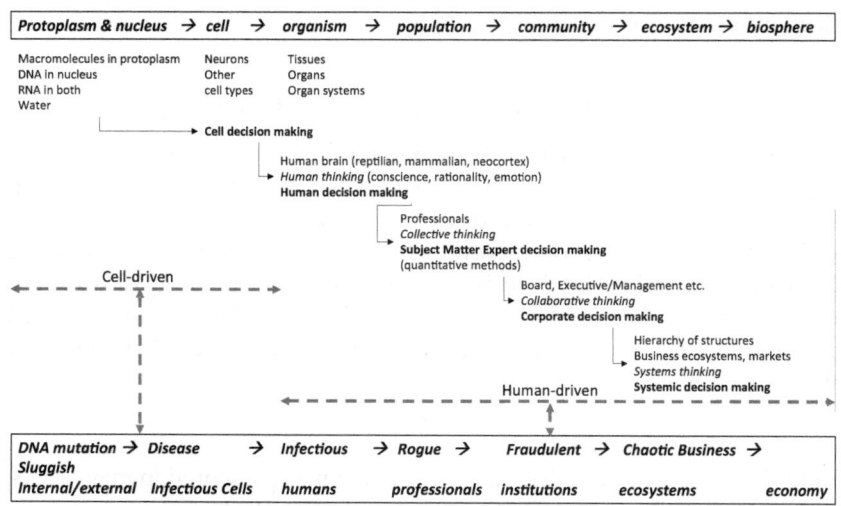

Fig. 5.1 Decision making in the biological spectrum

ness (alliances, suppliers, etc.). The market or economy also has a network like-structure, in the sense of Moore's business ecosystems or Barabasi's network (Barabasi et al., 2004).

Each of the five decision making schemes (ladder part of Fig. 5.1) will have its own, component-specific systems thinking to guide its decision making: cell decision making; human decision making; expert decision making; corporate decision making; economic decision making.

We take the view that corporate decisions are generated by human employees, individually or collectively. Although we are interested in rational decisions made via decision analysis and modern quantitative methods in decision science, we focus on irrational decisions investigated in decision neuroscience or neuroeconomics.

In the following we quickly review methods and practice of decisions. Then we present a systemic, biological spectrum-based approach to understanding, explaining, evaluating and predicting of decisions for appropriate outcomes towards corporate fiasco prevention.

On Decisions

A decision is commonly termed as good or bad, right or wrong, timely or untimely, and so on. Corporate decisions are taken by humans who run the institution (organization, corporation, firm, enterprise, etc.) in business, industry, or government, nationally or globally.

A Brief Literature Review on Decision Theory

Over many decades there have been different approaches to decision making creating theories, frameworks, processes, and/or techniques for different applications in different disciplines (Berger, 1993). We start with a quick overview of decision theories which belong to one or more of three domains: (1) middle level decision science developed for applications in institutions, markets and the economy; (2) lower level domain, called decision neuroscience; and (3) higher level domain of decision science or neuroeconomics .

Decision Science Approach
Traditional and modern decision theories are well presented in many texts, including Jim Berger or D. W. North (North, 1968). In our context, we review several decision development concepts, tracing back to Paul

Samuelson's decision-relevant developments from consumer choice and behavior in the market. Samuelson proposed the WARP (weak axiom of revealed preference) for preference choices, which was extended to GARP (generalized axiom of revealed preference).

A utility function then associated values to choices (or decision alternatives) and outcomes. Von Neumann & Morgenstern included the element of uncertainty in their expected utility (EU) theory (von Neumann et al., 2007). An EU over time, called discount utility, was added as discussed in Matta, Goncalves & Bizarro. Kahneman & Tversky with their prospect theory was next as an alternative to EU theory. Prospect theory looks at values assigned to gains or losses and decision weights rather than probabilities. Prospect theory thus extended the limitations of the EU approach (Tversky & Kahneman, 1981). Lars Skyttner presented an excellent treatment on decisions in his manuscript on General Systems Theory (Skyttner, 2005).

Decision Neuroscience Approach
At the neurophysiological level, according to Daeyaol Lee of the Kalvi foundation (Kavli Foundation, 2011), decision making involves the coordination of multiple brain areas. To understand lower level brain disorders and their effects, Lee took a top-down view to explore how prospect theory and reinforcement learning theory related to low-level brain decisions. Lee reviewed various functions used in decision making under risk in an economic context, such as a utility function for maximizing expected values, a value function, discount function and forgetting function for improved inter-temporal choices involving delays in rewards. He examined the shapes of the functions to determine the strength of risk, averse or seeking, in the context of prospect theory. Consequently, Lee provided an extensive review of decision making and the linkage between neurological and psychological levels, together called *cognitive neuroscience*. Any deviation from normal behavior would lead to different types of neurological and psychiatric disorders. This line of thought was discussed in Kable & Glimcher (Glimcher & Fehr, 2013) and through research at the Kavli Foundation.

In 2011 Kahneman developed a new line of thought called Think Fast or Think Slow (Kahneman, 2011). It consists of System I, which is limbic-driven (think fast), and System II, which is neocortex-driven (think slow). It further shows the psychological influence, linking economics to psychology, together called *behavioral economics*. The new trend is to look at neuroscience, psychology, and economics as an integrated neuroeconomics.

Other authors contributed to this newest approach, neuroeconomics. To expose high-level decisions involving low-level neuron activities, they developed experiments using fMRI and PET technology to decipher the neural activity of millions of neurons in the brain area and/or of a single neuron (applicable in experiments on animals) when humans made a high level decision.

Little attention, as far as we know, has been given to decision domains from the systemic perspective of von Bertalanffy-Boulding's general systems theory or GST, although there have been many application of GST in multiple disciplines. The important psychological foundation of George Kelly's Personal Construct Systems or PCT and its repertory grid (RG) technique have been explored for use in numerous disciplines, including business. However, they are kept separate from neuroscience.

We present a different approach to decisions as presented in the next section. We base our approach on the biological spectrum system ranging from protoplasm to biosphere for structuring the decision domain, which is parallel to but different from GST. Decision making is investigated from a system thinking approach, using PCT for problem solving towards fiasco prevention.

A SYSTEMIC APPROACH TO UNDERSTANDING DECISIONS

We add to Fig. 5.1 two features of human-driven decisions, networked and psychological, as an integrated neuroeconomics perspective of decisions and decision making.

From the human perspective there is a psychological influence, explored by considering human-driven decisions (shown in bottom dotted line). The psychological aspect of decisions will be explored for a better understanding of human decisions, using George Kelly's PCT and his RG technique under the umbrella of neuroeconomics decision making. Each decision making level is explained further below.

A cell consists of protoplasm (also called cytoplasm) and the nucleus. Protoplasm houses proteins and other components, while the nucleus contains DNA. RNA is located in both and is responsible for protein synthesis and messages between protoplasm and nucleus. The rest is water (75 %). There is communication between proteins within a cell and between cells resulting in what scientists call *cell decision making*, as described in Balazsi, van Oudenaarden & Collins (Balazsi et al., 2011).

At the next level (organism), the human brain consists of three brains in one, or triune (MaClean's model) (Newman et al., 2009): (i) the reptilian brain, the oldest brain and part of the subconscious mind; (2) the mammalian brain or limbic, which connects information to memory driven by emotion; and (iii) the neocortex, the newest brain where most thinking is done. The reptilian brain, consisting of the brainstem and the cerebellum, controls heart rate, breathing, body temperature, balance, and so on. It is compulsive and rigid, and is capable of making such decisions as flight or fight, and mating. As such, its lower-level contribution is more procedural, straight stimulus and response (unconditional or conditional).

The mammalian brain tends to drive humans to make decisions based on what and how they feel. The neocortex, primarily the prefrontal cortex and the precuneus, is believed to be overpowered by the older brains due to signals sent out by the amygdala. If decision makers are angry, they are more likely to be triggered by the reptilian brain. If they cry while thinking of a past event, they are influenced by the mammalian brain. If they consciously respond with logical thinking, they exercise the functionalities of the neocortex and precuneus.

It is argued that the decision to drop atomic bombs on Hiroshima and Nagasaki was not completely driven by sophisticated computational studies of all sides, Japanese, US, and even Russia. The decision included political studies, where and when to drop the two bombs (Little Boy or Fat Man), estimated casualties, and so no. It ended up, arguably, with President Truman's statement, "When you have to deal with a beast, you have to treat him as a beast", which revealed a decision more affected by the limbic system (Donahue, 2012).

The decision by Richard Fuld Jr. of Lehman Brothers to wait for the Fed to bail them out, as in the previous case of Bear Stearns in March 2008 was based on a guess, despite all prior negotiations to get Bank of America and Barclays Bank involved. And it did not happen.

The two examples above show elements (or factors) of emotion in the decisions. The complex nature of the human brain, its biology and psychology, has an effect on the final decision. The above are illustrated by Antonio Damasio's Descartes' error, showing that decisions are mostly emotional. With the above in mind, we present and discuss the model for understanding and evaluating decisions in the next section.

A CONCEPTUAL MODEL FOR UNDERSTANDING DECISIONS

Recall that from the perspective of corporate fiascos, we have sketched the following model of basic components in an institution (business, industry, or government): (i) the *institution* itself led by humans; (ii) the *exceptions* (errors, mistakes, wrongdoings, cover-ups, frauds, etc.); and (iii) *decisions* by humans which cause or fix exceptions (one way or another, good or bad). Multiple systems thinking on exceptions and decisions, as presented in the framework of Figs. 5.1 and 5.2, are exercised. These three components intertwine back and forth cyclically, as shown in Fig. 5.3.

In some organizations, when a crisis happens, a crisis management team is formed to handle it together with mainstream management. In our proposed model, we elevate the crisis management team to a permanent structure for enhanced management control and enhanced corporate governance. We call it the Oversight organization unit (to be detailed in Chap. 5).

We will look at the model from an initial perspective as clockwork at the human component level, within the larger context of either institution, market, or economy, as required. Information on every exception-decision pair is elicited. The human decision makers behind each pair of exceptions-decisions are interviewed for details. We used the RG technique as described below.

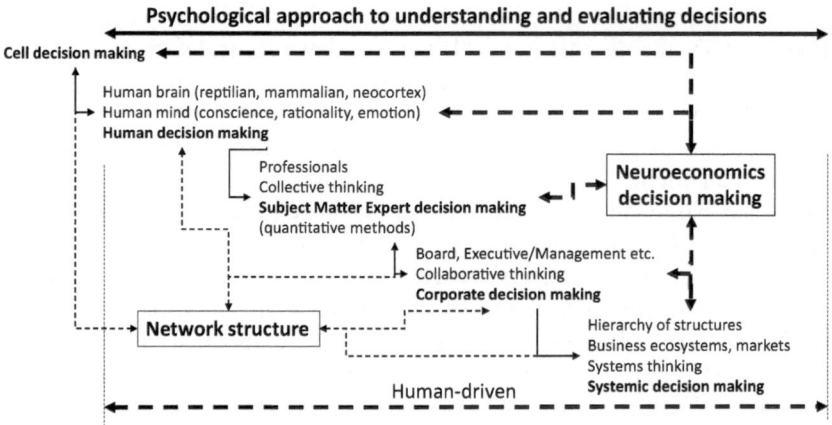

Fig. 5.2   Neurological-psychological approach

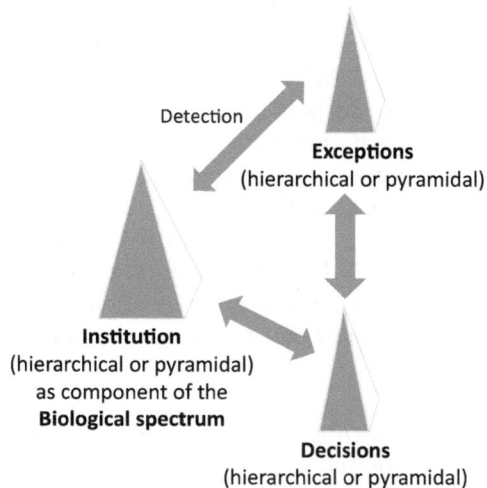

Detection

Exceptions
(hierarchical or pyramidal)

Institution
(hierarchical or pyramidal)
as component of the
Biological spectrum

Decisions
(hierarchical or pyramidal)

Fig. 5.3 Conceptual modeling

Understanding, Explaining, Evaluating, and Predicting Decisions

The key for understanding decisions is to elicit, analyze, and evaluate all the factors involved in a decision from the decision makers' perspective (i.e., to go inside their mind). We refer to George Kelly's famous work on PCT and RG technique used on his patients in a clinical environment. Our adaptation of RG is to extend the technique within the biological spectrum framework, which includes *biology, decision neuroscience, decision science, economics* and *psychology*.

There are two major tasks: (1) building a repertory grid; and (2) analyzing it to understand, explain (interpret), evaluate and predict the decision and its outcomes.

Building a Repertory Grid

First we, the interviewers, and/or the decision makers (DM) look at the problem at hand (the intertwined *exceptions* exposed and the resulting *decisions*) to identify the *elements* involved. The elements can be people, things or objects, activities or events, or other more specific business elements, such as projects, funding, accounts, policy, transactions, and so on.

Since the elements should be of the same *category*, as suggested by Stewart & Stewart (Stewart & Stewart, 1981), we consider multiple *categories* of elements, each with its own RG.

In the case of Barings, the people included: Leeson himself, his different managers, his employees, his customers, his counter partners in Tokyo Japan, SIMEX, the top executives in the UK, etc. The things or objects were: the futures in SIMEX or Nikkei 225, error account 88888, incentives, etc. The activities or events were: Kim Wong's initial loss of £20K, George Seow's loss of £100K, Kobe earthquake, the doubling scheme during January–February of 1995; there were also the decisions to conceal the losses in the error account, Barings audits, meetings with customers, etc.

Understanding and Measuring the Decisions Made on Exceptions

Decisions made can be viewed as a grid in Fig. 5.3. In this grid, we think of employees/SMEs as primarily concerned with operational, and some tactical, decisions in an institution. On the other hand, management leadership is primarily involved with strategic, and some tactical, decisions and with corporate vision.

Good management leadership and good employees/SMEs ensure success (quadrant 2 of Figure 5.4). This is in contrast and opposed to bad management leadership and bad employee/SMEs (quadrant 4). The latter could bring about business failure. If one of the two are bad while the other is good, it will be more complex to measure or label the failure/success (quadrant 1 or 3) of the decisions.

In all four cases (quadrants) we wish to know how to provide a measurement for decisions leading to the resulting failure/success measure of the institution/business. Besides Campbell's red flags in his discussion on why *good decision makers made bad decisions* (Campbell et al. 2009), we could consider the complex neurobiological view of a decision, or of a neurological view built on lower-level biological processes. Another could be an evolutionary biology view of decision making, which seems to be closer to a business aspect. We could also take a computational view of the decision process via Bayesian' reasoning, linear regression, optimization in decision theory, etc. But we propose a more practical approach, the top-down psychological aspect of decisions adapted from George Kelly's PCF with the RG technique.

Unfortunately, it is very difficult to use the technique for many reasons, including the fact that, first, we have to modify the technique for use

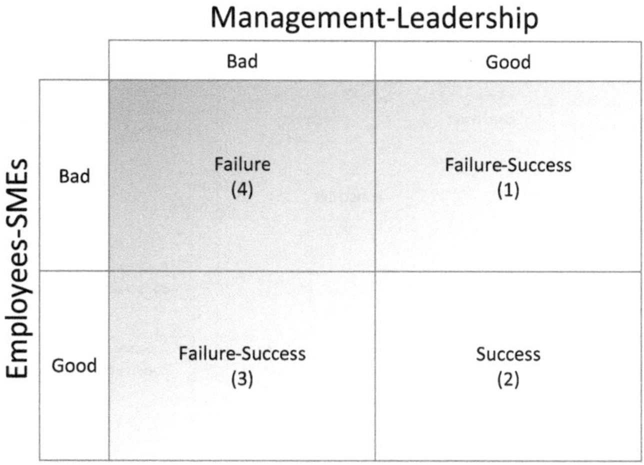

Management-Leadership

		Bad	Good
Employees-SMEs	**Bad**	Failure (4)	Failure-Success (1)
	Good	Failure-Success (3)	Success (2)

Fig. 5.4 Decision top grid of constructs

with our target decision makers. Second, and most important, just like the detection of red flags in Campbell's approach, it is highly unlikely that decision makers will reveal their model of the world, especially in cases where high-performance decision makers are unethical.

It is known that responsible parties do not always act properly or as expected on exceptions. At times, they are consciously involved in arbitrary decisions. They can intentionally ignore, avoid, alter, or hide those exceptions which could aggravate or lead to a fiasco. Therefore, it is necessary, but not sufficient, to expose exceptions and make them transparent. There must be a way to understand the decisions of responsible parties and to measure their decision effectiveness.

Repertory Grid of Decision Constructs: Elicitation, Analysis and Evaluation

Each decision maker (DM) has a mental map or model of success or failure in terms of his/her overall experience or in terms of a specific issue, in the sense of George Kelly. The DMs use their individual model consciously or unconsciously to decide what, when, with whom, and how to proceed, while prioritizing their daily tasks.

The process for arriving at a hierarchy of these constructs for analysis and evaluation is adapted from the RG technique, as in Fransella and other

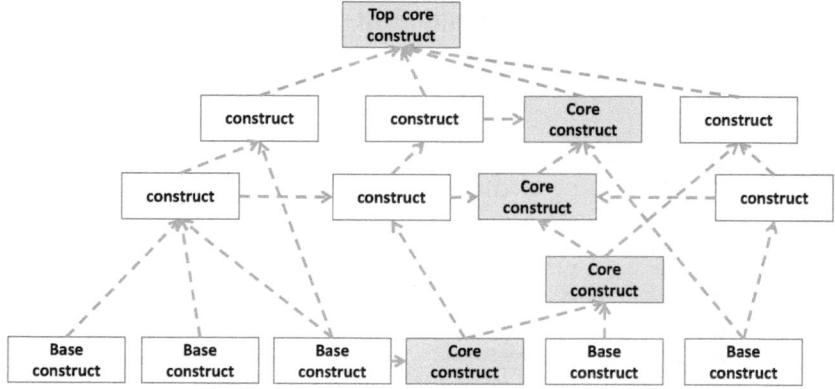

Fig. 5.5 Modified repertory grid

variations. Our purpose is to identify those core constructs which hurt the business (darker boxes of Figure 5.5).

Towards the top of the hierarchy are those more abstract constructs, such as mark to market (e.g., Enron's Jeff Skilling). At the bottom are the constructs which contribute to those exceptions which are indecomposable, such as a compensation amount paid to executives. In between the two, there might be constructs such as special purpose entities (SPEs) as used by Enron's Andrew Fastow.

Different to some other models, we assume that at each layer (or abstraction level), there is a core construct to be identified. In other words, we hypothesize that the linked core consists of several constructs which together influence the overall business stability.

Asking (Eliciting) the What

We, the interviewers, need to be unbiased, objective, and non-suggestive so that we can get the most out of the decision makers' minds. The RG suggests a selection of *elements*, of the same category as triads, and elicit their *constructs* (factors, properties, attributes of elements) via interviews. This same category criterion can be relaxed if elements of different categories make sense. As a starting point, we ask the DMs to select three elements at random from one category, (e.g., *people*). In the Barings example, the three elements are shown on the first row of Table 5.1, and marked with an "*".

Table 5.1 Repertory grid example

Similarity	Element1 Leeson	Element2 Customer	Element3 Employee	.... someone	ElementN Boss	Difference
Aggressive	(*) ✓		(*) ✓		(*) X	Relaxed
People-oriented	(*) X	(*) ✓			(*) ✓	Self-centered
Detailed Why: incentive	(*) ✓		(*) X	(*) ✓		Superficial Why: passive
...	...	...	...	...	...	...

We then ask the DMs

- *Which two of the three selected elements* have something in common (or similarity properties) in terms of how they feel about them in the context of the problem (exception or decision) at hand? We identify the similarity as a construct. For example, an answer is that Leeson and a particular employee are both *aggressive*. We place it in the column Similarity, and mark the two elements identified with a tick mark ("✓").
- We then ask in what way the two similar elements differ from the third. The decision makers come up with: *relaxed, passive, confined,* etc. as the opposite construct. We want the answer be specific. It cannot be "not aggressive" or "unaggressive" (that is not specific enough). In this example the opposite is that the boss is described as *relaxed*. The opposite is entered in the last column, and is marked in the grid as a cross mark ("X"). The initial results are as shown.

Asking (Eliciting) Why Using Hinkle's Laddering Up

For the dipole *similar-different* (or *similar-opposite*), we could ask the DMs to specify the one pole (*aggressive, relaxed*) they prefer to work with. For example, if the preferred pole is *aggressive* we have to find out why. There is a wide range of reasons why people are aggressive. Just like the selection of elements, or first level constructs, we want to really nail down the most relevant whys, in the instance of *aggressive* this might be the *financial incentives* or *plain fear of fraud being found out,* as in the Barings' Leeson case.

Hypothetically, take the reason for being *aggressive* is *incentive*, then the next question is what would be the opposite of *incentive*. We want to make sure the opposite of *incentive* is the why of *relaxed*, and this opposite makes sense. Otherwise we would have to investigate the proper reasons further. One possible opposite is *passive*.

At this point, we have two options: (i) continue to elicit the whys of *aggressive;* or (ii) abstract further by asking additional reasons *why incentive* or why the opposite, *passive*. Thus, we can have several first-level whys and more whys of the original whys, until we and the DMs both feel that the associated construct, *aggressive,* is detailed enough.

The process continues until all important constructs are addressed. Practically, Valerie Stewart suggests the number of first level constructs is around nine (Stewart, 2010). The number of whys for each why would be around five. Of course, after we embark on the analysis there is the possibility that the table (grid) will show some discrepancies, then we go back to the initial grid and make corrections.

Asking (Eliciting) How Using Landfield's Laddering Down
The purpose of laddering down is to find the *how*s concerning a similarity or difference. Take, for example, the construct *aggressive*. Aggressiveness can be active or passive. The how of being *aggressive* can be recognized by the way DMs yell at others on the phone when taking orders, pressure on traders, and other attitudes with an intention to cause harm. It could also be explored from the other pole, *relaxed*.

Filling in All Cells of the Grid
There are empty cells in the grid after all the first level constructs, and the why and how constructs are enumerated and placed at the two opposing ends, as observed in Table 5.1. They have to be filled in with tick and cross marks by the DMs. The DMs go to each cell and mark the empty cell appropriately. If an inconsistency is found at this time, or later, the DMs fix the entries in the cells. An example of a filled-in grid is shown in Table 5.2.

We explore all the answers to fill in each cell during the elicitation phase above within the context of the biological spectrum. It means that an answer can be at any component level: cell, human, institution, market, or economy. We, the interviewers, need to have a working knowledge on the three decision domains. For example, at the cell component level, we should be able to classify answers as originating from Maclean's model, following neural patterns of decisions, such as those described in Campbell, etc.

Table 5.2 Example of filled-in grid

Similarity	Element1 Leeson	Element2 Customer	Element3 Employee	.... someone	ElementN Boss	Difference
Aggressive	(*) ✓	✓	(*) ✓	X	(*) X	Relaxed
People-oriented	(*) X	(*) ✓	X	X	(*) ✓	Self-centered
Detailed	(*) ✓	✓	(*) X	(*) ✓	X	Superficial
Why: incentive	✓	✓	X	X	✓	Why: passive
How: yelling	X	✓	X	X	✓	How: silent
...	...	...	...	...	...	...

Table 5.3 Elements versus elements matrix example

Elements	Element1 Leeson	Element2 Customer	Element3 Employee	.... someone	ElementN Boss
Element1		2	7	4	8
Element2			5	6	4
Element3				1	7
...					...

Analyzing the Repertory Grid with a Consideration of Integrated Neuroeconomics

The methods for analyzing the grid are well documented in various sources. We only present a simple one here for illustration.

To start we compute the matrix *element by element*. In the two columns Element1 and Element2, if the row has the same tick or cross mark, we count it as 1, otherwise it is 0. In this example, the first row and third row both have (✓ and ✓) therefore a value of 1, the second row (X and ✓) therefore 0. For all three rows, the total is 2. We continue with two other elements to build a Elements versus Elements matrix, as shown in Table 5.3.

The value 8 in the first row indicates that the two Element1 and Element2 have the most similarities, as does the third row on Element3 and ElementN. This is an indication that the three elements (1, 3 and N) shown belong to some sort of a cluster of similarities. Therefore, when the grid is complete we can identify the clusters. The process is the same for other matrices, *Constructs versus Constructs*, and *Constructs versus Elements*. Also, instead of using the dichotomy scheme (✓ and X), we can use a ranking or a rating scheme (e.g., from 1 for total similarity to 5 for total difference).

On the surface, the repertory grid is simple to build and to compute. There are computer programs (e.g., Focus or Ingrid) available for preparing analyses by cluster analysis or principal component analysis (Bourne et al. 2005).

The elicitation and interpretation however are more difficult than we might think since we are attempting to go inside the mind of the DMs to find hidden facts on how they see their problems, environment, and world, and their logic. It requires some knowledge of the discipline in question, interview tactics, some working knowledge of decision science or neuroeconomics, as alluded to earlier. It also requires some characteristics of a psychologist, including patience, lack of bias, communications skills, emotional stability, ethics, interpersonal skills, open-mindedness, compassion.

Therefore, the process has to start small when the first significant exceptions occur. These interviewers, hereafter labeled as subject matter experts (SMEs), need to be trained sufficiently, much like the operations research analysts in the old days of MSOR. They will work on a regular basis alongside management, auditors, regulators, consultants, and other professionals in the institution. They should have full access to management by exceptions systems for the detection of exceptions. The exceptions should be transparent to all responsible parties.

Summary and Discussions

In this discussion, we have not actually conducted any interviews with decision makers in a critical situation. So we proceed somewhat differently without losing the generality of the proposed technique.

There are a couple of past examples under consideration: Nicholas Leeson of Barings Bank; Andrew Fastow and Jeff Skilling and their executive team at Enron; and Richard Fuld Jr. of Lehman Brothers. All three cases involved bankruptcy, costly impacts, and long-term ripple effects. We can investigate court records, congressional hearings, and hundreds of media and research articles on each case for information on how to capture the problems faced by decision makers from their perspective. Of course, this is not the same as an actual interview.

We chose the simplest case of the three: Leeson of Barings Bank. It turned out that this case revealed some interesting results, undetectable from Leeson's professional and/or social behavior. In this particular case, we did not have court records or congressional hearings.

In an attempt to understand Leeson's decision model we based our constructed scenario primarily on: Adams Curtis' 52 minute-documentary (Curtis, 1996); the book *Rogue Trader*; Rawnsley's *Going for Broke*; numerous media records; and Leeson's numerous interviews afterwards.

The Curtis documentary involved interviews with: Nicholas Leeson; his former wife Lisa; Peter Norris, CEO of Barings Investment Bank; Steve Clarke, Hong Kong Merchant Banker; Ron Baker; and Pamela Chiu. The interview format and questions did not follow the grid technique, but they helped find out partially why Leeson and Norris did what they did.

Rogue Trader, by Leeson himself, reflected on how he brought down the bank. Rawnsley retold the latest events which brought the bank to collapse. We thought they were adequate to illustrate our proposed technique, because our focus is on the technique itself.

Two-thirds of Leeson's book is on what happened and how it happened. Why he did what he did was partially explained in the remainder. Among the people he described were his employees, his superiors or people in supporting organizational units, his clients, the SIMEX authority and others, these fell into the *people* category. The error accounts, futures, options, margin payments, etc., constituted the category of *objects* he handled. Fraud, hidden schemes, meetings, audits, and so on were in the category of his *activities*.

The first thing we spot in the attempt to understand Leeson's model of his financial world is that Leeson was not in it for the money. He said he could have walked off with all the bearer bonds (equivalent to blank checks) from the Jakarta office, with a bundle of money, and never have to work again. He was ambitious. He was a go-getter. He left Morgan Stanley because he did not get the trader's job. He was very successful at his first job overseas in sorting out all the settlements transactions in Jakarta.

Leeson showed he cared about his employees, but in reality he was careless about them. He settled Kim Wong's errors because he hired her. She sold rather than bought 20 contracts, which yielded a loss of about £20K. He later bailed out George Seow, who bought rather than sold 100 contracts worth around £8M with a loss of around £150K. Both were recorded in the 88888 error account because he could not afford to report the losses, which would mean the death of his career.

He was manipulative in terms of people. He used them. He had a strong tendency for fraud. He found every which way to conceal his mistakes and those of others. He was a hard worker who liked pressure. He was quick to recognize holes in the system.

Leeson was a very private man. He did not share his thoughts fully with anyone, not even his wife, Lisa. He took extremely high-risks. He used Martingale's doubling scheme during February of 1995 in the hope that the Nikkei index would soon go up again after Kobe. The world was a game to him. In the end he did not blame anyone. He wrote a note, "I am sorry," and walked away.

Leeson was lucky in his job. He fixed Jakarta's problem and got recognized. He traveled with Tony Dickel to gain tremendous experience. He learned from everyone and anyone. He looked for loopholes and worked around systems. He took advantage of other people's ignorance.

We could see that his decision was based on the situation in front of him. He could have stopped the fraud scheme a couple of times, and did actually consider it. Circumstances drove him to success and to disaster. He committed fraud because the opportunity to do so was there. That was the way his brain was wired.

The thoughts and facts found in these documents can only be used for constructing partial repertory grids and is far from the collection of grids on different categories we were looking for (people, process, things, regulations, etc.), embedding the problem at hand. The way interviewers run the elicitation and analysis constitutes what we label as system thinking on fiasco prevention. It also turned out that the approach identified the need for an Oversight organization, if a scheme for prevention is conceptualized. We try to formalize the system thinking and address the Oversight unit in the next chapter.

This chapter addresses *decisions and decision making on exceptions* from the perspective of the *decision maker's model of the problem space in the environment* they are responsible for. It is based on the theory of personal constructs and the repertory grid technique, fathered by George Kelly, for the elicitation, analysis, and evaluation of decision constructs. It submitted a modified technique, which can be value-added to other computational approaches, such as decision analysis, expected utility theory, prospect theory, etc. As such, it could address the challenge of corporate and economic fiasco prevention. The development of our technique takes into account the functionality and behavior of the brain, as proposed by Paul

Maclean's three brains in one, discussed in Daniel Kahneman's thinking systems (system I and system II) and the neuroeconomic view of biology, psychology, and economics.

In summary:

- We presented an approach for *analyzing and interpreting decision and decision making process* based on George Kelly's Personal Construct Theory and the repertory grid. The grid is a two-dimensional matrix of elements and constructs of decisions which consider biological, psychological, and economic interpretations, in addition to computational mechanisms and methods widely exercised in decision science.
- We developed the model within a *systemic* framework based on a biological spectrum in nature. In this spectrum, we look for anomalies via analogies (similarities) and homeomorphism. This is different to Ludwig von Bertalanffy's General System Theory, or GST, for unity of science by isomorphism. It is also different from Kenneth Boulding's GST version describing the skeleton of science. To address the considerations of biology, psychology, and economics, the technique incorporates recent thoughts from neuroeconomics in recognizing the factors involved in the decision.
- We identified, in the process, the need for an Oversight organization unit. It will cost the institution in terms of creating the unit and running it with its own management team and SMEs as an opposite, independent, and parallel part of the whole organization. But the cost would be minimal compared to the cost of fiascos or bankruptcy, should the latter happen.
- The PCT of George Kelly is very enticing and his RP technique is easy to understand and apply. The reality in fiasco prevention is that it requires someone who is trained in decision science and is knowledgeable in neuroeconomics, a new category of subject matter expert.

In the big picture sketched in Fig. 5.4, we explore the analogies of exceptions between levels. We narrow down a practical model for addressing the issue of understanding decisions towards corporate fiasco prevention. We draw some insights and lessons learned from the literature to guide our proposed model (Fig. 5.5).

- The first is that we can use Paul Maclean's classification to initially identify the nature and type of a decision, e.g., when a decision maker angrily overrides a comment of his peer or subordinate, his logical thinking is overpowered by the limbic brain.
- The second is about the part emotion, according to Antonio Demasio, and the psychological aspect plays in decisions.
- The third is that we will not forget and/or downplay the importance of the well established and well used quantitative methods from Bayesian analysis to dynamic programming, and many other theories, such as prospect theory and beyond.
- The fourth is that we need to exercise system thinking across all decision domains, from neuroscience to neuroeconomics thinking, besides quantitative thinking.

A new perspective appears from discussions in the previous sections. Exceptions have to be monitored and double checked. Decisions can be questioned. But responsible parties might be blinded and take no action, as reported in Dharan & William. We need an Oversight organization unit, different from a common crisis management organization, to properly carry out monitoring and investigation tasks. This Oversight organization should be tasked by the board of directors with sufficient authority as shown in Fig. 5.2. This organization unit does not do strategy or consultation, just executes properly defined operational tasks and reports directly to the board of directors. Like a judge in the judiciary system, the Oversight organization personnel will be highly paid but receive no commission. It should be independent and staffed with its own SMEs, such as the interviewers we mentioned earlier.

Our approach adds value to the management of exceptions and understanding managerial or executive decisions. It would be an incremental step towards the prevention of corporate fiascos.

Fiasco prevention is a very complex and difficult task and our work just scratches the surface of this kind of problem. We hope it will start a series of discussions on the model leading to a prevention theory.

CHAPTER 6

Preventing Corporate Fiascos: *Corporate Oversight Organization Unit*

FORMATION OF AN OVERSIGHT ORGANIZATION UNIT

In Chap. 2, we have identified a series of issues after examining a couple of major cases: these were the Information issue, the Decision issue, and the Control issue. Chapter 3 was concerned with a systemic framework in which fiascos and potential solutions reside, and also concerned with a conceptual model of fiasco prevention. Chapter 4 covered the detection of information exceptions. In Chap. 4, we argued there were "disconnects" among different entities in the information system and elevated the functionality analogous to the autonomic system for the early detection, validation and exposition of symptoms. Chapter 5 related the understanding of decisions which responsible people made and challenged their appropriateness. Chapters 4 and 5 presented detailed discussions and examples on how they can be carried out.

But who would do the required tasks? Definitely, it would not be anyone in the line of command, that is the management and leadership teams and/or their employees. Definitely, it would not be Board of Directors, the accountants, legal counsel, the Ethics Committee, the Crisis Management Team or trades unions representatives. These organization units are too general or specific in their responsibility.

We conceive an Oversight Organization Unit (OU) which is permanent and parallel to the institution's management-leadership for enhanced corporate governance. It operates as a watchdog with capability for analysis and evaluation. It is chartered by the Board of Directors. It can

© The Editor(s) (if applicable) and The Author(s) 2016
T.N. Nguyen, *Preventing Corporate Fiascos*,
DOI 10.1057/978-1-137-49250-0_6

access independently corporate information management by exceptions and question management-leadership on decisions. It mimics partially the functionality of an immunization system in the cell-human component (Fig. 6.1).

How would the OU handle various situations: well, we have detailed some cases in Chaps. 2, 4, and 5? After the re-examination of the Barings and Enron cases, we thought the fiascos would not have occurred if Leeson's error account was questioned or Fastow's SPEs with the 3 % rule checked out. Since no one knew about this error account except his loyal and local subordinates, Leeson enjoyed exercising his dual role as general manager and trader, thus leading him to abuse of the system. Fastow enjoyed devising a violation without any objection from his accounting audit arm (which was also consulting), and therefore he got away with it. Were these people intelligent, or smart, or something else again? It appears that these violations originated from the character of the people in charge. Would the OU be able to stop them?

The next question was how would Leeson have been able to ask for some $30 to $40 million a day from Barings headquarters for margin calls without being officially questioned? How did Fastow continue to use SPEs in the thousands without any questions from Chairman Ken Lay of the Enron Board of Directors or from CEO Jeff Skilling? The OU involves as much in operations as in the line of command, so would it have been able to question Leeson's requests or Fastow's scheme?

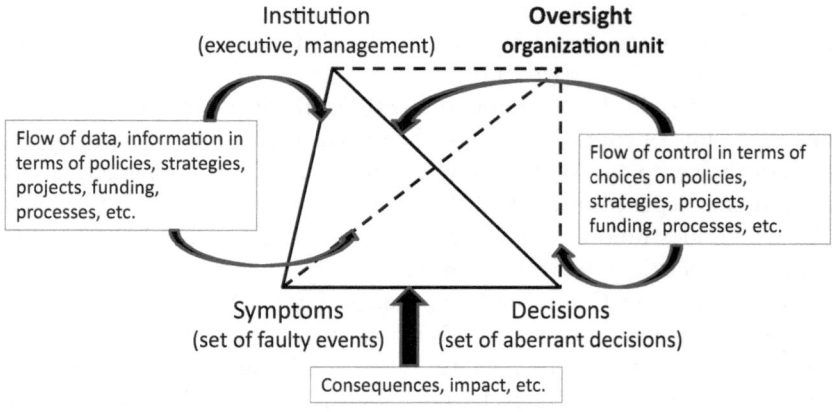

Fig. 6.1   Oversight Organization Unit (OU)

Apparently the Sarbanes–Oxley Act did not work completely. Reforms in accounting, in legal counseling, in corporate governance, did not stop fraud taking place. In fact, after the Barings fiasco, Jerome Kerviel of Societe Generale managed to perform unauthorized trading without his superior knowing it. The loss was huge. His activities went unnoticed. This case is different from Leeson in that Kerviel had back-office experience before he become a trader. The way he hid the loss was also different.

The story went like this. On January 18, 2008, suspicions were aroused about trading activity. Societe Generale was to investigate. It found that Kerviel had committed unauthorized trading activities and faked the records. A couple of days later, it reported the 4.9 Beuros (£3.7B) loss. Kerviel position was terminated. He was later charged with breach of trust, computer abuse and forgery. It was too late for Societe Generale as in the case of Leeson of Barings Bank.

In the Society Generale case, Axel Pierron, senior analyst at Celent, asked after the fraud was discovered: "How did someone who knew how the back office system work, become a trader?" He pointed out that since the trader knew very well how to circumvent the bank's risk-management system, he should not be a trader to segregate the back and front office. We disagree with Alex Pierron. We did not think that experience in another area, for example in back office operations, was a crime. The true issue was that we were dealing with people whose tendency was to commit fraud. Without the OU, this situation would be difficult to detect and stop.

The case of Lehman Brothers was a little different. Lehman Brothers was already deeply lacking of liquidity due to high leverage and the pursuit of a generally high-risk strategy. However, we thought that if the first Repo 105 was prevented from being reported as sales in the first place, the situation would not have happened. The damage would have been minimized.

The situation could have been avoided if, prior to the decision to use Repo 105, someone in the OU (not as a subordinate but as a watchdog authority) had officially challenged and/or helped deliberate the options. If Repo 105 was challenged, the issue of reporting loans as sales to a fictitious entity would have surfaced, and it would have been recognized as a fraud.

Actually, internally there were other alternatives even before the use of REPO 105 that Fuld did not consider. Instead, Fuld and his COO, Joe Gregory went for higher risk. Fuld instructed Mark Walsh to purchase Coeur Defense at $2.8B backed by MBS among other leverage buy-puts

(LBOs). Many of his top performers, VPs and managing directors, Mike Gelband, Christine Daley, Larry McCarthy and Alex Kirk, unable to persuade Fuld, have left. The VP who warned the Fuld–Gregory team about irregularities in accounting practices was fired. Bart McDade, replaced Gregory as COO in June 2007 and isolated Fuld and his team from the rest. He brought back Mike Gelband and Alex Kirk in the attempt to save the company, but it was too late. A OU could have intervened with decisions made by Fuld.

Fuld's final decisions were so bad and they led Lehman Brothers to bankruptcy. Richard Fuld decided not to consider Warren Buffet's initial offer. He did not take the offer made by the Korea Development Bank. What did he have in mind? Did he really think the Fed would bailout Lehman Brothers as it had Bear Stearns six months earlier? Did Fuld consul others? If he did, did the others understand and see the situation any differently? Would the situation have been different if he took other opinions and/or recommendations into account? The OU, in its chartered capacity, could have challenged Fuld's decisions.

Furthermore, corporate decisions during crisis were surprising. AIG top management, on the verge of asking for a bailout, had a corporate meeting for bonus decisions; the cost of such meeting ran in the hundreds of thousands of dollars. Other CEOs flew by private jet to meetings with potential rescuers. These activities were commonly practiced. The OU could have prevented these nonsensical activities.

The fact of the matter was that CEO Fuld, CEO Skilling and CFO Fastow were the top officers in charge. They could practically do anything they wanted, especially when they were driven by bonus incentives. Fuld walked away with a bonus in the tens to hundreds of millions each year. He had the *responsibility* to make this kind of decision. He knew he had the *authority*. We wondered if he spent any time thinking about his *accountability*. If he did, why did he take hundreds of millions in bonuses when his company was so vulnerable? This RAA (responsibility, authority, accountability) is crucial. The question is would an ad hoc ethics committee, as commonly existed in institutions, have prevented this wholesale fraudulent activity. The answer is perhaps and perhaps not! The chance for an ethics committee to be successful is slim because the committee members are commonly appointed by the CEO. Only an OU can challenge the CEO's RAA.

Most Fuld's decisions were emotion-driven. He was the one who at an internal meeting had this to say about his opponents: "*I reach in, rip out their heart, and eat it before they die"*. His decision had nothing to do with the complex combination of the *data* collected by the firm, *information* derived from it which could have been reported to him. His decision was not driven by his *knowledge* of the market and all players, or the experience he had gained over the years. Would the OU be able to bring him back to reality?

Also, there were problems reported in the existing corporate governance function in the institution in at least three bankruptcy cases: Barings, Enron and Lehman Brothers. Barings' top executives were in the dark. The accounting auditing function was given the responsibility of consulting (as with Enron). Ernst & Young failed to report any violation of standards at Lehman Brothers. In addition, Lehman Brothers' Board of Directors consisted of only ten members, and most of them were of the old school and did not know or really care about innovational techniques of 'making' money! Would a OU have played a correct role in fixing the situation?

In Chap. 2, we presented and characterized the fiasco problems. In Chap. 3, we suggested a systemic framework and a model for fiasco prevention. Chapters 4 and 5 exposed two main issues: exceptions and decisions. There is the management aspect of the two issues, communication and control, which Norbert Wiener calls 'cybernetics'. To explain this, we will use Wiener's own words: "When I communicate with another person, I impart a message to him, and when he communicates back with me he returns a related message which contains information primarily assessable to him and not to me. When I control the actions of another person, I communicate a message a message to him, and although this message is in the imperative mood, the technique of communication does not differ from that of a message of fact. Furthermore, if my control is to be effective I must take cognizance of any messages from him which may indicate that the order is understood and have been obeyed".

This cybernetics functionality is owned by the institution's leadership-management and/or its professionals along the line of command. The reality is that any decision might be hampered by the decision-maker's emotion, management ignorance or negligence, or top executive greed or risk, and the like.

We would like to duplicate this functionality and task it to an Oversight organization unit to exercise it from a different perspective. This would be an attempt to keep a degree of objectivity on viewing exceptions and implications of decisions. For example, in the case of Enron, Andrew Fastow decided to create a new SPE to handle pay back CALPERS's investment in JEDI I, and to lure CALPERS into invest in JEDI II. The Oversight organization unit would have needed to interview Andrew Fastow and his team. The OU would also interview the Arthur Anderson accounting auditing team and the legal team.

To be able to carry out its responsibility, the OU must have authority to deal with the management-leadership team. Thus the OU must be chartered by the Board of Directors. The OU can access fully the MBE and all decisions made by the management-leadership team. The OU would have its own subject-matter experts to carry out the interview for RG.

It was also reported that *cybernetics* has evolved since the over 1940s. Norbert Wiener originally coined the term to describe the union of principles in communication theory and control theory governing and directing a system. It was elevated from physiological and neurological feedback, together with prior work on homeostasis by Walter Cannon.

Mainstream economics (neoliberal economics) has been assessed as having been successful since the 1950s. The idea of economic cybernetics surfaced after the meltdown of the global financial system due to the subprime mortgage bubble in 2008. Researchers started to rethink and restudy the purpose and viability of mainstream economics in order to identify issues which led to failure. Within the framework set forth by Kennett Boulding, Robert Hoffman offered a new approach to economics called the 'cybernetic' approach.

Skills Set of the Oversight Organization Unit

At the onset, the OU should consist of personnel appointed by the Board of Director. It should exist in parallel to the management leadership team in the institution. It could be hierarchical or committee-based as appropriate. It could have its own subject-matter experts (SMEs). It primary responsibility should be to identify the signs of any disease in the institution, and inform the proper, responsible, parties, especially the Board of Directors. It should investigate signs and determine root causes, caused externally or resulting from internal decisions. It should be a partner of the leadership-management team. It should focus on management control and corporate governance.

Between 1939 and 1945 a new discipline emerged called Operations Research (OR). This discipline was concerned with highly technical theories, methods and tools for complex problem-solving. It used mathematical models, statistics and complex algorithms to find optimal or near-optimal solutions. It used optimization techniques (e.g. maximization of performance, or minimization of costs, risk, loss) over some objective functions to help top management to reach planned objectives. It was an interdisciplinary field. It was used to analyze strategic and tactical problems in military.

After the war, OR started to emerge to be employed in complex problems of business, industry and government, and in many different disciplines: production, finance, services, and so on. It had another hat called 'Management Science', and commonly referred to as MSOR.

People were trained as OR analysts acting as problem-solvers to help institutions achieve efficiency in operations and cost-effectiveness. They used large data, sophisticated modeling, computation techniques and analytics to help top management make strategic decisions. They have been employed in virtually all disciplines.

The OU should be staffed with these MSOR analysts. They should be trained somewhat in psychology and decision neuroscience and they certainly need to be knowledgeable on repertory grid techniques.

As such the OU will be independ and compepent to carry out its responsibility, authority and accountability in assisting corporate executives and managers in the line of command to do their job properly.

CHAPTER 7

Preventing Corporate Fiascos: *Beyond the Institutions*

Among the fiascos and collapses of institutions during the last two decades, Lehman Brothers' collapse in 2008 was of special interest to us as was one of the most important contributors to the financial crisis, as illustrated in Chap. 4. To face the financial crisis Fed Chairman Ben Bernanke and US Treasury Secretary Henry Paulson appeared in front of the House Committee to ask for funding to redirect the economy.

The Lehman Brothers case was different from the Enron case and much more serious. While Enron was a shock, the Lehman Brothers collapse was termed an earthquake or tsunami to the economic environment. It was the consequence of an unavoidable downturn in the subprime lending market.

Because of a lack of cash Lehman Brothers decided to cover up their losses from using Repo 105 and reported the loans using sales to a fictitious company as collateral. This was a fraud. We have examined in some detail the case of the Lehman Brothers fiasco in terms of decisions from the decision maker's perspective in the institution component in Chap. 5. The impact of this case and the ripple effects of this fiasco need further explanation.

The main causes of the Lehman Brothers bankruptcy, and those which preceded it (e.g. Bear Stearns) and those which followed it (e.g. General Motors, AIG, etc.) were two-fold: (1) subprime market crisis; and (2) derivatives. The first, the massive failure of the subprime lending market, was well described in Chomsisengphet & Pennington-Cross in 2006 (Chomsisengphet et al, 2006). The second was how Lehman Brothers failed in the use of derivatives.

© The Editor(s) (if applicable) and The Author(s) 2016   109
T.N. Nguyen, *Preventing Corporate Fiascos*,
DOI 10.1057/978-1-137-49250-0_7

We now elaborate on the two main activities to contribute to the whys of failure by the use of derivatives. One was the work and processes carried out by loan officers, brokers, securitizers, market makers, etc. The other was the MBS (Mortgage-based securities), CDO (Collateralized Debt Obligation), and CDS (Credit Default Swap) as further detailed below.

After a home contract is signed, the mortgage process starts. For a conventional loan or FHA loans, proofs of income are required, such as W2, employment record, and enough down payments, etc. For NINJA (No income, no job, asset) loans, the loan officers would convince buyers that no proof of income was needed. To reduce a monthly mortgage, loan officers would offer an adjustable rate (ARM) for a number of years. The borrowers understood that the rate would be much higher three or five years later.

They knew they would not be able to pay after the ARM expired, but it was explained that they had alternatives. First, they could always sell the house for a profit since housing prices, assumedly, would increase at a decent rate. Second, they could ask to refinance. Third, they could default the loan and walk away. A monthly mortgage was lower than rent anyway. They had nothing to lose. These, however, were very risky loans.

After thousands and thousands of loans of this type were completed, loan servicing began. Shadow banks and/or investment banks like Lehman Brothers and their mortgage arms—BNN or Aurora of Lehman Brothers—would have created MBS, CDO, and CDS as derivatives.

The sellers of these futures, forwards, options, or swaps, did not have to inform individual investors if the original mortgages were legal. This was the case of UBS in Switzerland. The banks sold the derivatives to investors, reported the sales as incomes, and transferred the risk of defaulted loans to the investors.

Everybody in the process benefited from a piece of this mortgage, as commission, compensation, bonus, or other incentives. The higher a person was in the institution, the more financial benefits they would receive. The scheme worked beautifully for everyone in the process, starting with the buyers. Only the investors now bore the risks.

The borrowers, on the other hand, did not know they would be responsible for capital gains, until they were faced with short sales and foreclosures. In these cases, someone in the process had to account for the increased equity of the homes, which was taxable. The situation was worse if the original home buyer had taken out a loan against the equity. Furthermore, to carry out the MBS or CDO, the banks needed to borrow money from somewhere to finance the derivatives.

In the end, the investors and other shadow banks might be short of cash when the CDS amounted to billions of dollars. Banks were closed. Investors panicked. Stock prices fell. The market faced a financial crisis.

To investigate the crisis beyond the institution level, in the following, we explore the analogy between cell-human and human-institution to beyond—i.e. institution-market and market-economy components—for insights into financial crises.

INSTITUTION-MARKET AND MARKET-ECONOMY COMPONENTS

At the cell-human component level, the ectoderm, endoderm, and meso-derm from the embryo differentiate during development into many differ-ent types of cells. Epithelial or neuron cells are derived from the ectoderm. Hormone secreting cells are derived from the endoderm. Blood cells and immune system cells from the mesoderm.

The cells make up tissues consisting of four types: epithelial, connective, muscle, and neuron. The tissues make up different organs (heart, lung, gut, etc.) and the organs make up ten different organ systems (circulatory, respiratory, etc.) in the human body. Thus, a hierarchical structure exists in the human body. There is another structure, the network structure as we can observe in the brain, where neurons are connected via synapses. The collection of all nucleus is called the gray matter and the collection of all dendrites-axons is called the white matter.

Institutions consist of humans from different professions (analogous to tissues), organized into basic functions: finance, accounting, human resources, management, legal, IS/IT/MIS, marketing, etc. (analogous to human organs). Most institutions organize these functions in a hierarchy, while others—such as the US Congress—are committee-based targeted at different domains (Armed Forces Committee, Budget Committee, etc.). A university, on the other hand, has both a hierarchical structure (the administration), and a committed-based structure at university, school or college, and department levels. This is different to the oganization of busi-ness and industry.

Observe that beyond the human-institution (i.e. in the institution-market and market-economy) components, the structure is more net-worked although, according to Ronald Coase, a hierarchy of structures exists (Coarse, 1937). The term *market* indicates a domain or medium where supply and demand for goods and/or services is engaged in terms of trading (buying or selling). Industry is about the production of goods and services for a market. Markets and industries can exist in sectors: finance, manufacturing, and so on.

The term *market* can also be seen as a *hierarchy of structures* (as in Ronald Coase), or *business ecosystems* (as in James Moore), or *network* as in Walter Powell (Powell, 1990), depending on how we look at it. Regardless of its structural organization, all entities in a market (whether resources, capital, partners, suppliers, customers, regulators, etc.) are linked (up, down, sideways, networked). The same kinds of links can be found in an economic sector—regional, national or global. Anything that happens to an entity will certainly affect others in the market, industry, or economy.

An *economy* involves all aspects of market and industry, whether production, distribution, or trade and consumption of goods and services. One can consider the economy—world, national or regional—as bounded by geography. In that sense, one might talk about the economy at the cell-human component level, to indicate supply and demand in the metabolic operations. The analogy between components is enticing.

Since normally everything works so well in the cell-human component, we might ask if there is any way we could mimic the functionality and processes in the cell-human component to achieve the same objective in a higher component—being *stability* in an institution, *equilibrium* in the market and *balance* in the economy.

From the perspective of fiasco, we have examined cancer as a deadly disease in the cell-human component with which to address fiasco in the human-institution, as we theorized in Chap. 3 with details argued in Chaps. 4 and 5. Thus, we could ask if there is any possibility of extending the *cancer* in human and *fiasco* in institution concepts to higher level components. We discuss this in the following.

Recall that in Chap. 3, we identified the supporting entities of the *cell-human* component as macromolecule, blood, cellular exchange, air, and DNA/genes. Macromolecules are of four types: primarily proteins to generate metabolism (enzymes, receptors, transport, etc.); carbohydrates for storage and structure; nucleic acids (information, transfer); and lipids (storage, insulation, membrane support).

Analogously, at the *human-institution* component all projects need funding, much like all macromolecules need air. The projects work with data made available for them, much like the macromolecules need blood. Transactions are what occur in an institution, much like the exchanges between cells. A well-defined project with adequate funding generates different transactions according to a well-defined policy and process to guarantee proper results. This is analogous to well-functioning macro-molecules receiving oxygen from the blood supply, generated by proper genes in the DNA.

We wish to extend the analogy between cell-human and human-institution to the *institution-market* (ecosystem) and *market-economy* (biosphere) components. We look at *contracts* (business deals) which exist between institutions in the market, which are considered analogous to projects in institution. Similarly, *investment* in the market-economy would be considered as analogous to contracts at the institution-market level.

While employees of an institution live in a *data environment*, institutions live in an *information environment*, and markets in a *financial environment*. By the same token, business transactions and business funding (at institution-market component level) and external transactions and hedging (at market-economy component level) can be considered as analogous to transactions and funding at the human-institution component level. Thus, all five supporting entities are listed as: (1) *market component*: contract, information, corporate transaction, hedging, and regulations; and (2) *economic component*: investors, finance, external transactions, investment, and law. Note that we can also identify the analogues at the guiding principle and organization layers, as sketched in Fig. 7.1.

To illustrate the analogy between institution and market, we can examine the Enron case. Fastow was the mastermind behind all SPEs. The SPEs themselves, allowed by SEC and GAAP, are not terribly difficult to understand. They consist of a general partnership, a limited partnership, and an independent outside investor with at least 3 % of total assets. The complexity of Fastow's SPEs was in the different financing schemes with banks, Enron stock guarantees, and terms and conditions for financing. They became convoluted SPEs with guarantees from all sources, primarily from Enron. It became difficult to trace who was doing what, when, and how financially.

An example is drawn from Benston & Hartgraves. Enron created Chewco to buy back $383M of CALPERS's interest in JEDI. To that end, Enron took $132M from JEDI, took a loan of $240M from Barclay, and $11.5M from an independent outsider. Fastow wanted to be the manager of the SPE. This would have violated the criterion of having an independent external investor, and the activity would have to be disclosed in a proxy statement, at the objection of Enron's outside legal counsel. Michael Kopper, a Fastow employee, was designated as manager, thus avoiding the proxy condition but still not observing the criterion of independent inves-

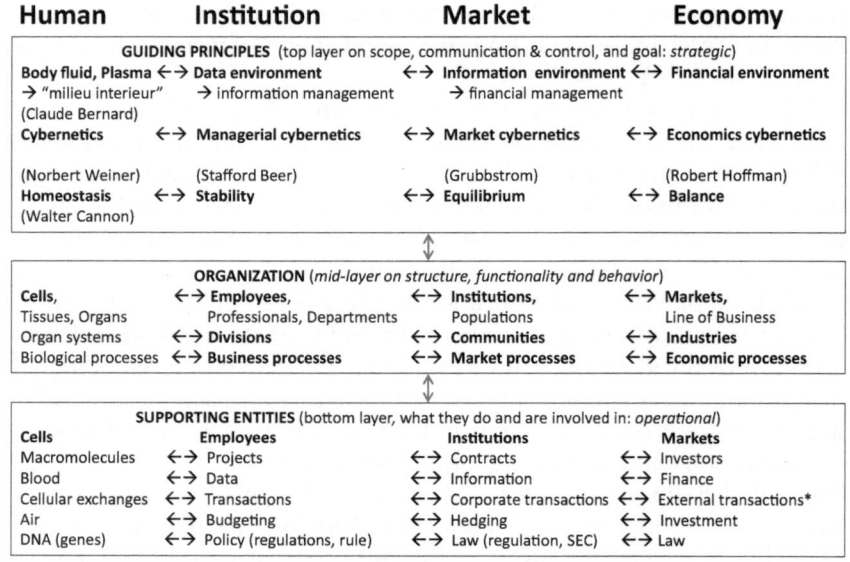

Fig. 7.1   Analogy between human, institution, market and economy

tor. Worse, of the $11.5M, most came from a Barclay's loan, with Kopper only putting down $125,000. One cannot understand how Fastow, a genius financier, could have committed to this scheme (Fowler, 2002).

This was a collection of SPEs. These included LJM Cayman (LJM1) which involved another SPE to hedge through LJM Swap Sub, and LJM Co-investment (LJM2) with four SPEs called Raptors, which created other SPEs. The purpose was to keep the losses out of the book.

Notice that all the SPE names were drawn from Star War characters or initials of Fastow's family members. This naming was not important to the financing scheme, but it did show that it was all a game for Fastow. His objective was to conceal losses if they occurred. The Raptor SPEs were to hedge a position in Rhythms NetConnections stock. He was thinking of power and control, of high risks, and of being above the law.

To illustrate the analogy between institution, market, and economy, we can also examine the Lehman Brothers case, in which the decision to enter the subprime market appeared normal. Later, however, Lehman Brothers became a victim of the subprime market crisis. Too many loans defaulted. Too many investors panicked and therefore withdrew. Banks refused credit lines or just stopped offering. Instead of making proper decisions, as recommended by managing directors and staff, Fuld did the reverse. He only listened to himself and his closest advisors. It was reported by MacDonald in 2008 that Fuld placed people without any ambition to replace him some day as his COO, such as Joe Gregory, just avoiding the possibility that someone would oust him, as Glucksman had ousted Peterson, Gluckman's mentor, in 1983.

Although we might label people such as Nicholas Leeson, Andrew Fastow, or Richard Fuld as abnormal people in the same sense that an abnormal cell causes cancer, one could say they are different to most. The problem was that they were not different in a good sense, with high ethics and good leadership. How come they were able to do what they wanted? The point was not only that these people believed what they wanted to believe, but also that others, such as Barings' leadership, did the same. Both Enron's and Lehman Brothers' boards of directors were incompetent and negligent, outside legal counsel and external accounting auditors were weak. Signs of anomalies in the institution surfaced, but they were ignored. The end result was an institutional collapse leading to its associated market crashing and to an economic crisis.

There has been a huge amount of literature on markets and economy, the two higher component levels in terms of monitoring market performance and national or global economic performance, market failure, and economic turmoil.

Our argument is that if institutions are stable, then the market is in equilibrium and consequently, the economy is balanced. The point is that one of the institutions could create a fiasco. What can be done then at the market and economy component levels? How can we monitor the market and the economy to detect exceptions and handle them properly to avoid market failure and/or economic turmoil? Would the operation of the five basic supporting entities help identify alternatives for the decision making by responsible parties using the same RG technique applied to them as to decision makers? After all, they are the movers, much like the institution decision makers aided by the Oversight organization unit.

Let us elaborate a little by taking a look at the Enron gas market and the WorldCom telecommunication market. In Chap. 2 we identified three issues after examining these two cases: an information exceptions issue; a decision issue; and an Oversight management issue. In Chap. 4 we argued that there were *disconnects* among entities in the institution and proposed a partial remedy for the detection, validation, and exposition of symptoms to eliminate the disconnects. In Chap. 5 we attempted to understand the decisions people made for correction if they were inappropriate. To guarantee the proper execution of the intertwined exception-decision complex, we introduce a model of Oversight organization in Chap. 6 and attempted to find out whether we can apply the analogous schemes to markets and the economy.

There are two issues: (1) would the guiding principles of cell-human component (*milieu interieur*, cybernetics and homeostasis), the organization (structural, functional, and behavioral) and the supporting entities (macromolecule, blood, cellular exchange, air, and genes-DNA) extendable to the market and economy components make sense, as sketched in Figs. 7.1 and 7.2; and (2) how would they fit in for further investigation?

The second issue refers to the process. At higher levels, these processes are more unstructured, as we might expect. Communication and control for market success depends on the environment within which the market exists.

Market Crash and Economic Meltdown Avoidance

The news media labeled the Lehman Brothers collapse as an earthquake or tsunami. Shell (Shell, 2015) reported six million jobs lost and a 5000 point plunged in DJIA. Lehman Brothers was the fourth largest investment bank in the USA. At the time of its collapse, its assets were $619B, with $639B in liabilities. It was the largest producer of subprime-based mortgage securities. Due to commercial MBS, Lehman Brothers was pressured to trade securities.

In most cases, we did not know the root cause of the collapse immediately after it occurred. Thanks to congressional hearings, news and media,

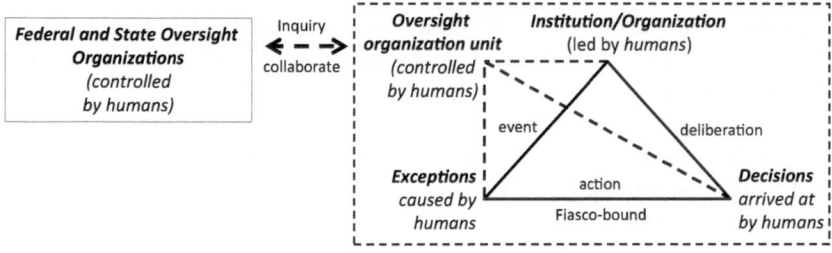

Fig. 7.2   Federal and State Oversight organizations

independent research, and court records we began to decipher them. Recall that, in our theme, we claim that most important root cause is decision makers who are in a position to move the business, strategically or operationally. All the characters we have examined are such people. It is their decisions, combined with their emotions in the presence of exceptions, that have had an effect. The exceptions can be the wants or the needs, different to expected normal goals or objectives. Hank Paulson Jr. refused to bail out Lehman Brothers and cited that his hands were tied because the Fed did not have sufficient funds. Paulson also said that he advised Fuld to find buyers multiple times. Lehman Brothers could not. There was a report which revealed that potential buyers would not receive the same guarantee as others. Apparently, the personal relationship between Paulson and Fuld was not that great.

We did not know about the fraud in Lehman Brothers until Anton Valukas completed his report in 2010. However the impact of Lehman Brothers' collapse was immediate. Several money funds, the Bank of New York Mellon and Prime Reserve Fund, fell below $1 per share. The impact propagated to Putman Investments, Evergreen Investments, Royal Bank of Scotland Group, major banks in Japan and in Hong Kong. They were linked to Lehman Brothers via bonds and loans facing billions of claims against Lehman Brothers. Freddie Mac's exposure in terms of principal and accrued interest was also in the order of billions. Stocks of other companies were severely exposed by Lehman Brothers' collapse.

The question to ask is, would the tetrahedron model have been able to prevent the Lehman Brothers fiasco (Valukas, 2010), and others, so none of the above would have happened? Following Bear Stearns collapse, would a modified model have reduced the exposure of others? The answer is possibly as shown in Fig. 7.2. In this modified version, existing federal and state Oversight organizations would have inquired into the institution's Oversight organization unit for conducting an understanding of the institution decisions, as we have described in Chap. 5.

There is a possibility that the federal and state Oversight agencies need an extension of current regulations to handle this situation, if one does not already exist. We propose that when symptoms and signs surface, as in the case of Lehman Brothers, the Oversight organization (federal, state or professional) ought to be involved. Rather than a stress test, as in the case of NY Fed which imposed on Lehman Brothers, Morgan Stanley, etc. these organizations or agencies would ask the institution's Oversight unit to investigate.

CHAPTER 8

Epilogue

TOWARDS A MATHEMATICAL BASIS FOR FIASCO PREVENTION

In the prologue, a systemic approach to corporate fiasco prevention was introduced. It is descriptively systemic, for two reasons: (1) the framework is under the umbrella of the biological spectrum; and (2) systems thinking is explored in each of its components as well as across components. The approach was empirically-based however, because it was populated with past case analyses of corporate fiascos (Chap. 1) to get insights into the conceptual model (Chap. 2).

The conceptual model was derived using case materials derived from Barings Bank, the Enron Corporation and Lehman Brothers. Chaps. 3, 4 and 5 detailed the components of the model. The key concept was the *exception-decision* complex. Exception is everything or anything which goes wrong, expressed as detectable signs and symptoms. Corporate decisions regarding the exceptions were measured numerically by the neurological-psychological repertory grid technique. The control of this set of all exception-decision complexes is assigned to an *Oversight* organization unit.

The model is to make sure business stability is maintained, analogous to maintaining homeostasis in a human body. The final chapter, Chap. 6, was concerned with the analogical extension to the market and economic crises as the institutions are components of the market within the economy.

© The Editor(s) (if applicable) and The Author(s) 2016
T.N. Nguyen, *Preventing Corporate Fiascos*,
DOI 10.1057/978-1-137-49250-0_8

The model implies that, if institutions are stable, the market and the economy would be also. However, it does not guarantee that chaos or turmoil do not occur if the causes originate from the higher components and affect the component parts downstream. Forthermore, to be complete and sound a quest for a theory of 'prevention' which is mathematically sound should be considered.

Is there a theory of prevention? As far as we know, not yet. We do have prevention methods or prevention practice, however. In the medical arena, vaccination is a method of preventing diseases. There is prevention of substance abuse. There is preventive dental practice. There are multiple methods proposed for crime prevention. It is not hard to find preventive methods in other disciplines; however, a mathematically sound treatment of prevention does not yet exist. We would like to propose a theory of prevention which encompasses fiascos in institutions, chaos in markets, and turmoil in the economy. The following argues the why, what and how of such a theory.

The Why Towards a Theory of Prevention

Insights from the Laws of Nature

The biological spectrum can be extended to the natural spectrum with inclusion of particles in the microscopic world on the left (see Fig. E.1) and the transcendental systems, i.e. the unknowables of von Bertalanffy-Boulding, in the cosmic world on the right. Thus all of the components will obey the laws of the nature, namely the Newton law on force and gravitation, the Coulomb law on charge, the Faraday law on induction, the Maxwell law on electromagnetic field, the Planck law on quantum, and the Einstein law of relativity. The exception-decision complex should be governed by those laws.

Just as a discrete phenomenon is a subset or a special case of continuous phenomena, the natural spectrum is part of the natural continuum in which everything is either created by nature or by living organisms, of which the human species is the most intelligent one.

The question is: "*Is it possible that the laws (from Newton's laws to the Einstein theory of relativity) shed any insights into the quest for understanding the human condition (employees and employers), institutions, markets and economies and dealing with fiascos within the context of the exception-decision complex environment?*" After all, decisions can be conceived as driving forces, influences, and/or inherent energies that affect the environment.

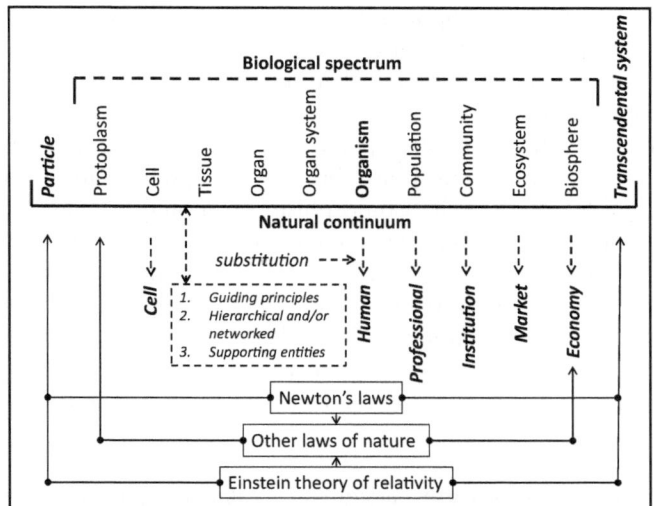

Fig. E.1 Extension towards a theory of prevention

Insights from Homeomorphism Between Components
There is a fundamental issue between components. The lower-level components in the biological spectrum such as the human cell are better understood than many higher-level components. The higher-level components are more abstract, and therefore the more unstructured and complex they become. Recall that the cell-human component was used as the base analogue in previous chapters.

In fact, we have exploited analogy concepts and applicable analogical reasoning between different component levels, primarily between the human cell and the human institution from the base analogue. In addition, within the analogy between two components, from top (general) to bottom (specific), the analogy between *top-layer guiding principles* is identified and appears clearer. The analogy between mid-layer *organizations* in terms of structure, functionality and behavior is less clear among components due to the differences in terms of how they address the environment, at each analogous component level. The analogy between bottom-layer *supporting entities* on the operations of each component is the least comparable due to details, the reason being that at the lower level the more specific it is in meeting structural, functional and behavioral requirements.

When macromolecules in human-cell components are equated to projects in human-institution components, then to contracts (business deals) in institution-market components and finally to investment in market-economy components, the analogy weakens as we go up component level since the next higher level is more than the sum of its parts There exists an element of emergence.

Nevertheless, there is another possibility which one needs to investigate fully, which is the homeomorphism between components in the biological spectrum, different than the isomorphism promoted by von Bertalanffy. The homeomorphisms between components are based on known processes in a known lower-level component such as human-cell or human-institution, for application to higher components such as institution-market or market-economy components. The idea is that further detailed investigation of one lower component can lead to those in the higher level component by virtue of topological invariants between them.

Insights from σ-Field

With the numerical measures, assigned or calculated, associated with the exception-decision complex, as hinted in the Prologue and detailed in other chapters, the *σ-field* on the *exception-decision* complex set emerges. With σ-field, let us define an exception-decision *impact* e_d as the difference between μ_i and μ_k of exception-decision i and exception-decision k respectively or ($e_d = \mu_k - \mu_i$). The difference can be positive or negative. To avoid dealing with large values of e_d, e_d is normalized using $f(e_d) = e_d / (1 + | e_d |)$. The range of normalized e_d will be $[-1, 1]$.

One possible application of σ-field to corporate fiasco prediction is to evaluate the sequence or subsequence of impact $\{e_d\}$ over time. If the sequence is non-increasing, it will converge and there will be a local equilibrium. If not, that is it is non-decreasing, it will diverge. If divergent, a subsequence should be injected into the sequence to alter the divergence.

If the exception-decision complex value is qualitative (or categorical), its value can be *no impact* $e_d = 0$, *highly negative* impact $e_d = -1$, *negatively impact* with $e_d = -1/2$, *highly positive impact* $e_d = 1$, or *positively impact* $e_d = \frac{1}{2}$. Thus, the sequence for consideration on convergence or divergence will be a sequence of values taken from the set of $\{-1, -1/2, 0, 1/2, 1\}$. The elements of the sequence can be in any order, therefore the sequence or the series (i.e. sequence and the sum of sequence) can either convert or diverge.

The good news is that given a divergent sequence or series, one can inject a subsequence, which is a controlled one, to the current sequence to force it converge. It implies that the subsequence, when known, can be translated into a set of decisions to prevent crisis. Of course, it is easy said than done.

The What to Be Considered in a Theory of Prevention

Identifying DNA of Higher Level Components
On the one hand, since a human being is determined by DNA as chain of peptides consisting of A, C, G, T nucleotides (Alberts et al., 1998), one would think there should or must be an analogous DNA for institutions, a DNA for markets, or a DNA for economies. There have been attempts to identify them as corporate culture analogous to human personality. Gary L. Neilson, Bruce A. Pasternack, and Decio Mendes of Booz Allen Hamilton suggested structures, decision rights, motivators, and so on as organizational DNA for institution (Neilson et al., 2004).

Identifying Complex Organizational Structure of Higher Components
On the other hand, of recognizable importance is the fact that a human being has a brain with a mind, unlike any other species, whose ability is practically boundless. The human mind brings us anywhere, any place, any time in the past and in the future. In fact, we only need to think of any place we have been, the mind and memory will bring us there. Thus with the ability of the mind we can, individually or collectively, recollect experiences in the past and project strategic plans for the long-term future across some space in a fashion similar to the gravitational space-time curvature of Einstein's theory of relativity.

One should recognize that some important concepts have been exploited and explained in the institution, market and economy component in a normal working environment. They are the *nature of the firm*, according to Ronald Coase, which led to Coase's theorem on external transaction costs. There was also the concept of *business ecosystem*, defined and explained by James Moore, when the market is considered from the ecological aspect. There was also the discussion of the *market as a network mesh*, suggested by Williamson, much like the neurons in the human brain.

Since the biological spectrum includes the market as ecosystem, and the economy as biosphere, these concepts should fall into its scope. The RG technique for understanding decision makers at every component level requires extensive knowledge and skills in operations, management, finance, accounting, regulations, market, and economy. It might require new laws and regulations, extended scope of responsibility for the *Oversight* organizations, and professional associations in terms of additional reforms, The latter would be viewed from a systemic aspect, so that disconnects can be avoided.

Characterizing Problems in the Context of Laws of Nature, Topological Space and σ-Field

Let us review the fiasco due to decisions of Peter Norris, CEO of Barings Investment Bank. When he decided to explore securities opportunity in South East Asia, he could never have thought that in less than three years later, Nicholas Leeson, a general manger of Settlements of Barings Futures, would single-handedly bring the oldest bank of the UK to its knees. Good opportunity does not always bring success.

Another one was President Bill Clinton. President Clinton would never have thought that investment banks have taken advantage of the law to act inappropriately, leading to the 2008 financial crisis. He signed the bill by *Gramm-Leach-Bliley repealing Glass-Steagall* Act of 1933 into law in 1999 to allow cross-subsidies and interconnections between commercial and investment banks. President George W.Bush signed the American Dream Act, intended to assist *low-income and low-scored credit* Americans with the opportunity to own a home.

President Bush would never have thought that someday a NINJA (**no income, no job or asset**) candidate would be able to purchase a home. Not only a few NINJAs, but there were millions and millions of them, whose mortgages were sold in the form of Mortgage-based Securities (MBS) and Collateralized Debt Obligation (CDO) to investors.

When these NINJA mortgages, with adjustable rate mortgage (ARM in the order of 2–2.5%) in the subprime lending market defaulted due to much higher rate (in the order of 7 or 8%) than the initial ARM rates, they led New Century or Bear Stearns to be short of cash, then to collapse, General Motors to bankruptcy, and so on. One of the largest investment banks on Wall Street, Lehman Brothers, was also out of cash, with liabilities amounted to above the value of its assets. Lehman Brothers had to file for 'Chapter 11' bankruptcy after all attempts at rescue failed.

The examples above point out that a good and hopeful decision might lead to an extremely bad outcome, *fiasco*. It implied that fiasco did not just happen. Somehow along the path from A to Z, some relatively small and abnormal event occurs. If not identified and acted upon in time, they could cause fiascos.

In the case of Barings, the small event started with a honest mistake by Kim Wong, a new trader, which resulted in £20K loss. She bought rather than sold derivatives. Leeson, her manager, reported it to Simon Jones, his boss. Leeson was advised to report it to Barings Tokyo. He did not. He decided to hide the loss in an error account named 88888.

Simon Jones did not care to follow up. Three days later the loss amounted to £60K due to a drop in price. Leeson continued to hide the subsequent loss and many others due to his poor trading activities. Worse, to balance the Profit and Loss statement, Leeson faked the gains. Barings Headquarters in London did not understand futures and options trading. They seem to be satisfied with the balance sheet with faked profits. Everyone in Barings was after yearly bonuses brought in by Leeson and his traders' activities. Management failed to check Leeson properly. Peter Baring, the Chairman, even said it was not that difficult to make money from futures arbitrage.

So, a fiasco results from a series of exceptions-decisions, one after another. The associated decisions or, at times, "no decisions", are mostly generated from the top leaders and their lieutenants. In the case of Barings, leadership did not do anything significant. They just watched with little attention to why some initial warnings were brought up. They let Leeson play the whole game of deception and fraud. After the collapse of Barings, Peter Norris, reportedly, said that he wanted to punch Leeson in the face. That was the acknowledgement of responsibility from Peter Norris, CEO of the Barings Bank. The picture was terribly wrong.

In the case of Lehman Brothers, there was a great pool of talent from vice-presidents and managing directors down, with excellent traders and researchers providing all possible information. They met every week on early Tuesday morning assessing the moves in the financial world in many areas. They knew their business. Yet, Richard Fuld Jr., CEO of Lehman Brothers kept himself somehow disconnected from the rest of his organization. He was advised to make the bank's position as short as possible, to downsize and recover. Fuld did not listen. He continued to involve in multiple leverage buyouts (LBOs) in addition to huge loss of CDOs. Lehman Brothers went bankrupt. In both cases, it was all about hugely incorrect executive decisions.

In fact, what the institutions, the markets they are in, and the associated economy will face in the future, inevitably, will be some newly human-created fiascos. They might be caused by some misunderstood, misguided, incomplete, undetected information, etc., and/or by some aberrant decisions. If these decisions are thought through and carried out properly at all times, Peter Norris would not have wanted to punch Leeson in the face, President Clinton's objective would have been met and his vision would have led to a balanced, affordable and happy housing environment.

Finding Mathematical Solutions

It should be stressed that there is no doubt that fiascos that occur in any institution can be extremely complex. It is still very difficult fully to understand a particular fiasco even after the facts are gathered, analyzed and discussed, especially when the root cause was fraud. Most responsible parties took the Fifth Amendment during congressional hearings. We need a systematic scheme for addressing corporate fiascos and beyond with a mathematical approach.

The How to Proceed in a Theory of Prevention

To address a theory of prevention, there are two parts:

1. The first is the formulation of the σ-field on a set of entities with a numerical measure μ, subjectively estimated as shown in Chap. 3. The concept of *entity* is used to present everything or anything in the world, abstract or real. Three entities, *institution, exception* and *decision,* need particular attention. *Institution* is an open system defined by Ludwig von Bertallanffy. *Exception* refers to any activity or event that goes wrong. *Decision* is any idea or action causing the exception or fixing it. There will be an σ-field on the exception-decision complex on which a complete algebra is defined.
2. The second is the topological mappings defined in the *σ-field on exception-decision complex. The decisions can be* expressed as the force applied to and/or energy generated on the components (institution, market or economy) in the economic space-time curvature. The understanding of decision impact would help idenfity path for preventing and/or anticipating crisis.

Since all entities are parts of the natural continuum, the solution surely exists within the system itself.

The above suggests a collection of concepts to be quantified and measured, which could be further developed based on insights from biology, physics, psychology, neuroscience, decision science, abstract algebra and algebraic topology into business operations and management. With these insights, we would go deeper via analogy and analogical reasoning, as well as further exploit the laws of nature and mathematical tools for detailed specifications of the problem towards a theory of prevention.

BIBLIOGRAPHY

Ackman, D. (2002). Enron's lawyers: Eyes wide shut? *Forbes.* http://www.forbes.com/2002/01/28/0128veenron.html

Alberts, B., Bray, D., Johnson, A., Lewis, J., Raff, M., Roberts, K., & Walter, P. (1998). *Essential cell biology.* New York: Garland Publishing.

Alexander, C. (2002). *The nature of order.* Berkeley: The Center for Environmental Structure.

Azadinamin, A. (2012). The bankruptcy of Lehman Brothers: Causes of failures & recommendations going forwards. *Swiss Management Center.* http://ssrn.com/abstract=2016892

Balazsi, G., van Oudenaarden, A., & Collins, J. J. (2011). Cellular decision making and biological noises: From microbes to mammals. *Cell, 144,* 910–925.

Bannister, D., & Mair, J. M. M. (1968). *The evaluation of personal constructs.* London: Academic Press.

Barabasi, A., & Oltvai, Z. N. (2004). Network biology: Understanding the cell functional organization. *Nature Review, 5,* 101–113.

Beer, A. S. (1972). *Brain of the firm: Managerial cybernetics of organization.* London: Penguin Press.

Berger, J. O. (1993). *Statistical design theory and Bayesian analysis* (Springer series in statistics). New York: Springer.

Bierman, H., Jr. (2008). *Accounting/finance lessons on Enron—A case study.* Singapore: World Scientific Publishing Co. Pte. Ltd.

Boulding, K. E. (1956). General system theory—The skeleton of science. *Management Science, 2*(2), 197–208.

Bourne, H., & Jenkins, M. (2005). Eliciting managers' personal values: An adaptation of the laddering interview method. *Organizational Research Methods, 8,* 410.

Brickey, K. F. (2003). From Enron to WorldCom and beyond: Life and crime after Sarbanes-Oxley. *Washington University Law Quarterly, 81*, 357–402.

Brown, S. J., & Steenbeek, O. W. (2008). Doubling: Nick Leeson's trading strategy. *Pacific-Basin Financial Journal.* http://pages.stern.nyu.edu/~sbrown/leeson.PDF

Bush. G. (2003). American Dream Downpayment Act. https://georgewbush-whitehouse.archives.gov/news/releases/2003/12/20031216-9.html. December, 2003.

Camerer, C. F. (2013). A review essay about foundations of neuroeconomics analysis by Paul Glimcher. *Journal of Economic Literature, 51*(4), 1155–1182.

Campbell, A., Whitehead, J., & Finkelstein, S. (2009). Why good leaders make bad decisions. *Harvard Business Review, 87*(2), 60–66.

Cannon, W. (1963). *The wisdom of the body.* New York: The Norton Library, Norton & Company.

Celani, C. (2004). The story behind Parmalat's bankruptcy. *Executive Intelligence Review, 31*(2), 10–12.

Chatterjee Sayan, & Batten Fellow. (2002). Enron's asset-light strategy: Why it went astray. http://citeseerx.ist.psu.edu/viewdoc/download?doi=10.1.1.197.9682&rep=rep1&type=pdf

Chomsisengphet, S., & Pennington-Cross, A. (2006). The evolution of the subprime mortgage market. *Federal Reserve Bank of St. Louis Review, 88*(1), 31–56.

Clayton, R. J., Scroggins, W., & Westley, C. (2002). Enron: Market exploitation and correction. *Financial Decisions, 14*, 1–15. Article 1.

Clinton, B. (1999). Gramm-Leach-Bliley Act, http://www.presidency.ucsb.edu/ws/?pid=56922 , November 12, 1999.

CNNMoney. (2002). Andersen auditor questioned Enron. http://money.cnn.com/2002/04/02/news/companies/andersen_bass/

Coase, R. H. (1937). The nature of the firm. *Economica.* New Series, *4*(16), 386–405.

Congressional Testimony-1. (2002). Sherron Watkins. http://www.sherronwatkins.com/sherronwatkins/Congressional_Testimony.html

Congressional Testimony-2. (2002, February 7). CEO Jeffrey Skilling's testimony to Congress.

Curtis, A. (1996). Nick Leeson and the Fall of the House of Barings, https://www.youtube.com/watch?v=NUaCo_bPePw&list=PL7RYGizYx40ml2Osjkf CMt8euee5rieAQ&index=1, 1996.

Damasio, A. (2005). *Descartes' error: Emotion, reason and the human brain.* New York: Penguin Book.

Deakin, S., & Konzelmann, S. J. (2003). Learning from Enron. *ESRC Centre for Business Research.* University of Cambridge.

Desmond, M. R. (2008, September 9). Lehman's Brothers. *Forbes.* http://www.forbes.com/2008/09/16/lehman-layoffs-biz-wall-cx_mrd_0916lehmanbar.html

Dharan, B. (2002). Enron's accounting issues—What we can learn to prevent future prepared testimony. *US House Energy and Commerce Committee's Hearing on Enron Accounting.* http://www.ruf.rice.edu/~bala/files/dharan_testimony_enron_accounting.pdf

Dharan, B., & Rapoport, N. (Eds.). (2009). *Enron and other corporate fiascos.* Eagan: Foundation Press.

Dharan, B., & William, R. (2002). Red flag in Enron reporting of revenues and key financial measures. http://www.ruf.rice.edu/~bala/files/dharan-bufkins_enron_red_flags.pdf

Donahue, N. (2012). Understanding the decision to drop the bomb on Hiroshima and Nagasaki. *Center for Strategic & International Studies.* http://csis.org

Easterby-Smith, M. (1980). The design, analysis and interpretation of repertory grid. *International Journal of Man-Machine Studies, 13,* 3–24.

Edwards, F. R. (2003, October). US corporate governance: What went wrong and can it be fixed? *BIS and Federal Reserve Bank of Chicago conference.*

Eichenwald, K. (2005). *Conspiracy of fools: A true story.* New York: Broadway.

Eichenwald, K., & Brick, M. (2002). Enron collapse: The strategy. *NY Times.*

Eisenhardt, K. M. (1989). Agency theory: An assessment and review. *Academy of Management Review, 14*(1), 57–74.

EnronAnnualReport1998. (1998). picker.uchicago.edu/Enron/EnronAnnualReport1998.pdf

EnronAnnualReport1999. (1999). picker.uchicago.edu/Enron/EnronAnnualReport1999.pdf

EnronAnnualReport2000. (2000). picker.uchicago.edu/Enron/EnronAnnualReport2000.pdf

Foster School of Business, Washington. (2010, April). A decade's worst financial scandals.http://foster.uw.edu/research-brief/the-decades-worst-financial-scandals/

Fowler, T. (2002). LJM2 partnership files for bankruptcy, Unit created to hide debt in Enron deals. *Houston Chronicle.*

Fox, L. (2003). *Enron: The rise and fall.* Hoboken: Wiley.

Fransella, F., & Bannister, D. (1977). *A manual for repertory grid technique.* London: Academic Press.

Free, C., Stein, M., & Macintosh, N. (2007, July/August). Management controls: The organization fraud triangle of leadership, culture and control in Enron. *The Organization.* Ivey Business Journal, July-August 2007.

Fromm, M. (2004). *Introduction to the repertory grid interview.* Münster: Waxmann Verlag.

GAAP. (2002). The Enron fraud why didn't anyone see it? http://www.thegaap.net/articles/The_Enron_Fraud.html

Gale. (2012). *Corporate disasters: What went wrong and why.* Detroit: Gale, Cengage Learning.

GAO-2009. (2009). Review of future combat system is critical to program's direction. http://www.gao.gov/new.items/d08638t.pdf

GAO-2012. (2012). 12-565R: DOD financial management: Reported status of Department of Defense's enterprise resource planning systems. http://www.gao.gov/products/GAO-12-565R

Glimcher, P. W. (2003). *Decision, uncertainty and the brain: The science of neuroeconomics* (A Bradford book). Cambridge, MA: MIT Press.

Glimcher, P. W., & Fehr, E. (2013). *Neuroeconomics: Decision making and the brain.* London: Elsevier.

Glimcher, P. W., & Rustichini, A. (2004). Neuroeconomics: The consilience of brian and decision. *Science, 306,* 447–452.

Glimcher, P. W., Camerer, C. F., Fehr, E., & Poldrack, R. A. (2013). Introduction: A brief history of neuroeconomics. In *Neuroeconomics: Decision making and the brain.* London: Elsevier.

Gregory, D. (2014). *Unmasking financial psychopaths: Inside the minds of investors in the twenty-first century.* New York: Palgrave Macmillan.

Greenspan Hearing, (2010), https://www.youtube.com/watch?v=0atvIK6l1TU, Financial Crisis Inquiry Commission, April 7, 2010

Gross, C. G. (1996). Claude Bernard and the constancy of the internal environment. *The Neuroscientist, 4*(1), 380–385.

Haglund, A. (2011). Failed strategies of Enron. http://mg312.wordpress.com/2011/09/02/failed-strategies-of-enron/

Heally, P. M., & Palepu, K. G. (2003). The fall of Enron. *Journal of Economic Perspectives, 17*(2), 3–26.

Heller, R. (2003, December 30). Parmalat: A particularly Italian scandal. *Forbes.* http://www.forbes.com/2003/12/30/cz_rh_1230parmalat.html

Higgs, D. (2003). Review of the role and effectiveness of non-executive directors. *Department for Business, Enterprise and Regulatory Reform.* http://www.berr.gov.uk/files/file23012.pdf

Hoffman, R. (2010). A cybernetic approach to economics. *Cybernetics and Human Knowing, 17*(4), 89–97.

Hogen, W. P. (1996). The Barings collapse: Explanations and implications. *IDEAS.* http://ideas.repec.org/p/syd/wpaper/2123-6743.html

Holtzman, M. P., Venuti, E., & Fonfeder, R. (2003). Enron and the raptors. *The CPA Journal.* http://www.nysscpa.org/cpajournal/2003/0403/features/f042403.htm

Iansiti, M., & Levien, R. (2003). The new operational dynamics of business ecosystems: Implications for policy, operations and technology strategy. *Harvard Business School,* Working paper 03–030.

IESE Professors. (2013). The tremendous effects of Lehman Brothers' fall five years later. *Economics.* http://blog.iese.edu/economics/2013/09/17/the-tremendous-effects-of-lehman-brothers-fall-five-years-later/

Iwata, E. (2006, January). Enron's legacy: Scandal marked turning point. *USA TODAY.* http://usatoday30.usatoday.com/money/industries/energy/2006-01-29-enron-legacy-usat_x.htm

Jickling, M. (2003). The Enron collapse: An overview of financial issues, RS21135. Updated 30 Jan 2003, *CRS Report for Congress.*

Kable, J. W., & Glimcher, P. W. (2009). The neurobiology of decision: Consensus and controversy. *Neuron, 63,* 733–745.

Kahneman, D. (2011). *Thinking fast and slow.* New York: Farrar, Strauss and Giroux.

Kahneman, D., & Tversky, A. (1979). Prospect theory: An analysis of decision under risk. *Econometrica, 47*(2), 263–291.

Kanaracus, C. (2012). The scariest software project horror stories of 2012. http://www.cio.com/article/print/721628

Kast, F. E., & Rosenzweig, J. E. (1972). General system theory: Applications for organization and management. *Academy of Management Journal, 15*(4), 447.

Kavli Foundation. (2011). The neuroscience of decision making. http://www.kavlifoundation.org/science-spotlights/neuroscience-of-decision-making

Kelly, G. (1963). *A theory of personality: The psychology of personal, constructs.* New York: W. W. Norton & Company.

Ketz, J. E. (2003). *Hidden financial risk: Understanding off-balance sheet accounting.* New York: Wiley.

Kim, S. (2011). 10 things we didn't learn from Enron scandal. http://abcnews.go.com/Business/10-things-learn-enron-scandal-10-years/story?id=15048641

King, R. J. B. (1996). *Cancer biology.* Harlow: Longman.

Kirkpatrick, G. (2009). The corporate governance lessons from the financial crisis. *Financial Market Trends,* OECD, *1,* 61–87.

Kolmogorov, A. (1956). Foundations of the Theory of Probability (2nd ed.). New York: Chelsea.

Korenini, B. (2012). Conducting consisting laddering interviews using CLAD. *Metodološki zvezki, 9*(2), 155–174.

La Roche, J. (2011). And now we know why nobody took Madoff whistleblower Harry Markopolos seriously… http://articles.businessinsider.com/2011-10-07/wall_street/30253397_1_trial-dates-bny-mellon-bank

Landau, M., & Chrisholm, D. (1995). The arrogance of optimism: Notes on failure-avoidance management. *Journal of Contingencies and Crisis Management, 3*(2), 67–80.

Le Maux, J., & Morin, D. (2011). Black and white and red all over: Lehman Brothers' inevitable bankruptcy splashed across its financial statements. *International Journal of Business and Social Sciences, 2*(20), 39–65.

Leber, J. (2012, August). How to spot the next big banking scandal. *MIT Technology Review.* https://www.technologyreview.com/s/428734/how-to-spot-the-next-big-banking-scandal/

Lee, D. (2013). Decision making: From neuroscience to psychiatry. *Neuron, 78*, 233–248.

Leeson, N. (2012). *Rogue trader*. Boston: Little Brown Book Group.

Li, Y. (2010, October). The case analysis of the scandal of Enron. *International Journal of Business and Management, 5*(10), www.ccsenet.org/ijbm

Looijen, R. C. (2009). Holism and reductionism in biology and ecology: The mutual dependence of higher and lower level research programmes. Ph.D. thesis. Rijksuniversiteit Gromningen.

Ravn-Jonsen, L. (2009) : Ecosystem management: A managementview, Working Paper, Department of Environmental and Business Economics, University of Southern Denmark, No. 86.

Lorsch, J. W., & Lawrence, P. R. (Eds.). (1970). *Studies in organization design.* Homewood: Richard D. Irwin.

Lyke, B., & Jicking, M. (2002, August). WorldCom: The accounting scandal. *CRS Report for Congress*, RS21253.

Mack, T. (1993). Hidden risks. *Forbes, 151*, 54–56.

Matta, A., Goncalves, F. L., & Bizarro, L. (2012). Delay discounting: Concepts and measures. *Psychology & Neuroscience, 5*(2), 135.

Maturana, H. R., & Varela, F. J. (1980). *Autopoesis and cognition: The realization of the living*. Dordrecht: D. Reidal Publishing Company.

McRoberts, F. (2002, September). The fall of Andersen. http://www.chicagotribune.com/news/chi-0209010315sep01-story.html#page=1

Michel, N. (2013, September 12). Lehman Brothers bankruptcy and the financial crisis: Lessons learned. *Issue Brief. The Heritage Foundation Leadership for America*. No. 4044.

MIT Sloan Newsroom. (2002). Can we stop another Enron? Kothari says new laws won't. *MIT Sloan Newsroom*. http://mitsloan.mit.edu/newsroom/2002-kothari.php

Moore, J. F. (1993). Predators and prey: A new ecology of competition. *Harvard Business Review, 71*(3), 75–86.

Moore, J. F. (1996). *The death of competition—Leadership and strategy in the age of business ecosystems* (p. 297). New York: Harper Business.

Moore, J. F. (1998). The rise of a new corporate form. *The Washington Quarterly, 21*(1), 167–181.

Nakayama, A. (2002). Lessons from the Enron scandal, http://www.scu.edu/ethics/publications/ethicalperspectives/enronlessons.html

Newman, J. D., & Harris, J. C. (2009). The scientific contribution of Paul D. Maclean. *Journal of Nervous and Mental Disease, 197*(1), 3–5.

Neilson, G., Pasternack, B., Mendes, D. & Tan, E. (2004). Profiles in Organizational DNA Research and Remedies, http://www.strategy-business.com/article/rr00004, February 4, 2004

Nguyen, T. (2007, March). Cancerous enterprises (Presentation). *CMU/SEI, SEPG 2007*, Austin Texas.

Nguyen, T. (2011, August). People, process, technology and IS: A shift of perspective for success. *Proceedings of the Shanghai international conference on social science*, Shanghai.

Nguyen, T. (2013, July). Cancer and deadly infectious diseases in institutions: Developing use cases for an MBE application to preventing another Enron or Barings. *Proceedings of the ICCGI*, Nice.

Nguyen, T. (2014). A different approach to information management by exceptions: Towards bankruptcy prevention. *Information & Management, 51*(1), 93–103.

Nguyen, T. (2015, January). Biological systems thinking for business. *Proceedings of the Business Systems Lab*, Perugia.

Nguyen, T., & Cvetojevic, V. (2015). On the basis of a computational model for decision making. *Proceedings of WDSI* (Western Decision Science Institute) 2010, Lake Tahoe, April 2010*.

Nguyen, T., & Dimitrov, D. (2012, May). An approach to institutional disease prevention. *Proceedings of annual conference on global economics, business and finance*, Beijing.

Nguyen, T., & Tran, K. (2014, April). A measure of management-leadership using a failure-success scale. *Proceedings of the WDSI conference*.

North, D. W. (1968). A tutorial to decision theory. *IEEE Transactions on Systems Science & Cybernatics, SSC-4*(3), 211–219.

Nuland, S. (1997). *The wisdom of the body*. New York: Alfred A Knoff.

NYTimes. Text of Watkins' letter to lay after departure of chief executive. *NY Times*. http://www.nytimes.com/2002/01/16/business/16TEXT.html

Oppel, R. A., Jr. (2002, May 8). Enron's many strands: The strategies. *NY Times*.

Patsuris, P. (2002, August 26). The corporate scandal sheet. *Forbes*. http://www.forbes.com/2002/07/25/accountingtracker.html

Patten, D. B. (2002, June). Summary of the Enron trading strategies in California. http://www.potomaceconomics.com/uploads/midwest_presentations/trading%20strategies%20assessment.pdf

Paulsen, S. (2002, January). Workers lose jobs, health care and savings at Enron. http://www.wsws.org/en/articles/2002/01/enro-j14.html

Peck, J. (2010). Memorandum decisions. http://www.nysb.uscourts.gov/sites/default/files/opinions/173916_89_opinion.pdf

Petrick, J. A., & Scherer, R. F. (2003). The Enron scandal and the neglect of management integrity capacity. *American Journal of Business*. http://www.emeraldinsight.com/doi/abs/10.1108/19355181200300003

Peursem, K. V., Zhou, M., Flood, T., & Buttimore, J. (2007, June). Three cases of corporate fraud: An audit perspective. Number 94, *University of Waikato*.

Pfarrer, M. D., Decelles, K. A., & Smith, K. G. (2008). After the fall: Reintegrating the corrupt organization. *Academy of Management Review, 33*(3), 730–749.

Pirtošek, Z., Georgiev, D., & Gregorič-Kramberger, M. (2009). Decision making and the brain: Neurologists' view. *Interdisciplinary Description of Complex Systems, 7*(2), 38–53. Scientific Journal.

Pollock, E. J. (2002). Enron's lawyers faulted deals but failed to blow the whistle. *The Wall Street Journal.* http://faculty.msb.edu/...ching/ENRON/ Enron's%20Lawyers%20Faulted%20Deals%20But%20Failed%20to%20 Blow%20the%20Whistle.htm

Powell, W. W. (1990). Neither market non hierarchy: Network forms of organization. *Research in Organizational Behavior, 12,* 295–336.

Powers, W., Jr. (2002). *Report of Investigation.* http://picker.uchicago.edu/ Enron/PowersReport(2-2-02).pdf

Puscas, D. (2002). A guide to the Enron collapse: A few points for a clearer understanding. *Polaris Institute.* www.polarisinstitute.org

Raghavan, A. (2002, May). How a bright star at Andersen burned out along with Enron. *Wall Street Journal.*http://www.wsj.com/articles/ SB1021425497254672480

Rapoport, A. (1968). *General systems theory, Systems science and cybernetics, I,* EOLSS.

Rawnsley, J. (1995). *Going for broke.* London: Harpercollins, 1e.

Reilly, S. (2012). How the air force blew $1B in a dud system. *Federal Times.* http://www.federaltimes.com/article/20121126/DEPARTMENTS01/ 311260009/How-Air-Force-blew-1-billion-dud-system

Reinstein, A., & Weirich, T. R. (2002). Accounting issues at Enron. *The CPA Journal, 72*(12), 20–25.

Ricketts, J. A., & Nelson, R. R. (1987). Management-by-exception reporting: An empirical investigation. *Information & Management, 12*(5), 235–246.

Rogers, M. (2012). Sarbanes-Oxley 10 years later: Boards are still the problem. *Forbes.* http://www.forbes.com/sites/frederickallen/2012/07/29/sarbanes-oxley-10-years-later-boards-are-still-the-problem/

Schadler, T. (2013). Healthcare. Gov's failure starts with leadership, not technology.http://www.forbes.com/sites/forrester/2013/10/25/healthcare-govs-failure-starts-with-leadership-not-technology/

SEC-1. (2004). SEC versus Richard A. Causey. http://www.sec.gov/litigation/ complaints/comp18551.htm

SEC-2. (2004). SEC versus Jeffrey K. Skilling and Richard A. Causey. http:// www.sec.gov/litigation/complaints/comp18582.htm

Senate Committee on Governmental Affairs. (2002).The role of the board of directors in Enron's collapse. https://www.gpo.gov/fdsys/pkg/CPRT-107SPRT80393/pdf/CPRT-107SPRT80393.pdf

Shadlen, M. N., Kiani, R., Hanks, T. D., & Churchland, A. K. (2008). Neurobiology of decision making: An intentional framework, in better than conscious? In C. Engel & W. Singer (Eds.), *Decision making, the human mind, and implications for institutions.* Cambridge: MIT Press.

Shell, A. (2015). Lehman Bros. collapse triggered economic turmoil. http://abc-news.go.com/Business/lehman-bros-collapse-triggered-economic-turmoil/story?id=8543352

Skyttner, L. General Systems Theory: Problems, Perspectives, Practice (2Nd Edition) 2nd Edition,World Scientific Publishing, 2005.

Smith, N. C., & Quirk, M. (2004). From grace to disgrace: The rise and fall of Arthur Andersen. *Journal of Business Ethics Education, 1*(1), 93–131.

Snowden, D. J., & Boone, M. E. (2007). A leader's framework for decision making. *Harvard Business Review, 85*(11), 69–76.

SOX. (2002). *Sarbanes-Oxley Act.* http://www.soxlaw.com

Sridharan, U. V., Dickes, L., & Caines, W. R. (2002). The social impact of business failure: Enron. *Mid-American Journal of Business, 17*(2). http://www.bsu.edu/mcobwin/ajb/?p=199

Starling, E. (1923). *The wisdom of the body.* London: H. K. Lewis.

Stein, M. (2000). The risk taker as shadow: A psychoanalytic view of the collapse of Barings Bank. *Journal of Management Studies, 37*, 8.

Stevens, J. R. (2008). The evolutionary biology of decision making. *Faculty Publication*, Department of Psychology, Paper 523. http://digitalcommons.unl.edu/psychfacpub/523

Stevenson, R. W. (1995, February 27). Markets shaken as the British bank takes a big loss. *NY Times.*

Stewart, V. (2010). http://valeriestewart-repertorygrid.blogspot.com/

Stewart, V., & Stewart, A. (1981). *Business applications with repertory grid.* London: McGraw Hill.

The Enron Blog. (2008). *Jonathan Weil's six business lessons.* http://caraellison.wordpress.com/2008/06/12/jonathan-weils-six-business-lessons/

Thomas, C. W. (2002). The rise and fall of Enron. *Journal of Accountancy, 193*(4), 41–54.

Touryabal, H. (2012). 10 biggest banking scandals of 2012. *Forbes.*http://www.forbes.com/sites/halahtouryalai/2012/12/27/10-biggest-banking-scandals-of-2012/#1a5eea6dd7da

Tversky, A., & Kahneman, D. (1981). The framing of decisions and the psychology of choice. *Science.* New Series, *211*(4881), 453–458.

US GAAP. (2002). The Enron fraud why didn't anyone see it? http://www.thegaap.net/articles/The_Enron_Fraud.html

US versus Andrew Fastow. (2003). www.dindloa.com

Valukas, A. (2010). Lehman Brothers Holdings Inc. Chapter 11 Proceedings Examiner's Report http://web.stanford.edu/~jbulow/Lehmandocs/origIndex.html, March 2010

von Bertalanffy, L. V. (1950). An outline of general system theory. http://www.isnature.org/Events/2009/Summer/r/Bertalanffy1950-GST_Outline_SELECT.pdf

von Neumann, J., & Morgenstern, O. (2007). *The theory of games and economic behavior*. Princeton: Princeton University Press.

Watkins, S. (2001). Letter to Ken Lay. www.findlaw.com, news.findlaw.com/hdocs/docs/enron/empltr2lay82001.pdf

Wiener, N. (1948). *Cybernetics or control and communication in the animal and the machine*. Cambridge: The Technology Press.

Williamson, O. E. (1992). Markets, hierarchies and the modern corporation: An unfolding perspective. *Journal of Economic Behavior and Organization, 17*, 335–352.

Zimmerman, P. B. (2012, March). Decision making for leaders in proceedings advanced leadership think tank series. *Harvard Business Review*. http://advancedleadership.harvard.edu/files/ali/files/decision_making_thinktank_final.pdf?m=1412019787

INDEX

© The Editor(s) (if applicable) and The Author(s) 2016
T.N. Nguyen, *Preventing Corporate Fiascos*,
DOI 10.1057/978-1-137-49250-0